AF531250

LEARN & TEACH
VEDIC MATHEMATICS

Published by
Lotus Press Publishers & Distributors

LEARN & TEACH

VEDIC MATHEMATICS

Cosmic Kapoor, S.K.

Ved Rattan

Lotus

PRESS

4735/22, Prakash Deep Building,
Ansari Road, Daryaganj,
New Delhi - 110002

Lotus Press Publishers & Distributors
Unit No.220, 2nd Floor, 4735/22, Prakash Deep Building,
Ansari Road, Daryaganj, New Delhi- 110002
Ph:- 32903912, 23280047, 098118-38000
Email : lotuspress1984@gmail.com
Visit us : www.lotuspress.co.in

Learn and Teach Vedic Mathematics

ISBN : 81-89093-00-2

Published by : **Lotus Press Publishers & Distributors,** New Delhi- 2
Printed at : **Bharat Offset Works,** New Delhi

Foreword

Dr Kapoor's approach to Vedic Mathematics is a geometric one, based on the structure of the Vedas and the Vedic literature. This therefore, offers a new approach to the teaching of mathematics, which has many important and profound implications. Though children deserve and expect a holistic education modern teaching remains fragmented and lacks coherence, so the unified approach to mathematics teaching outlined in this book and based on Vedic texts is very welcome.

Dr Kapoor has demonstrated that the organization of knowledge in the Vedic literature mirrors the structure of Vedic Geometry. This Vedic Geometry transcends our familiar three-space geometry as it includes higher-dimensional spaces; all spaces having a four-fold structure. These are real spaces, which can be experienced by transcending the lower-dimensional spaces. The practice and understanding of transcendence is vital in education. Higher states of consciousness, which correspond to the experience of higher-dimensional spaces, need to be included in any educational system, so that children recognise what they are already aware of: that reality goes beyond the familiar world of three-dimensional space.

Every discipline should be taught in a way that brings out its full range and potential. The Ganita Sutras expounded by Sri Bharati Krsna Tirthaji cover all of mathematics, both pure and applied, and as Dr Kapoor has shown, these have their own geometric format. Dr Kapoor's approach to the teaching of mathematics is therefore a coherent and integrated one, which can transform mathematics education into a powerful tool for individual development. But first teachers must learn the Vedic system and how to teach it, and to this end Dr Kapoor has given courses on learning and teaching Vedic Mathematics which are currently being made freely available on the Internet.

Kenneth R. Williams

Foreword

Preface

While sitting in a room, one gets conditioned as if the universe is just 3-Space. The movement one comes out of the room, the one with melting of the conditions enjoys the bliss as the universe being a 4-Space.

Soon this blissful state manifests as a mental block and one becomes conditioned as caged within 4-Space.

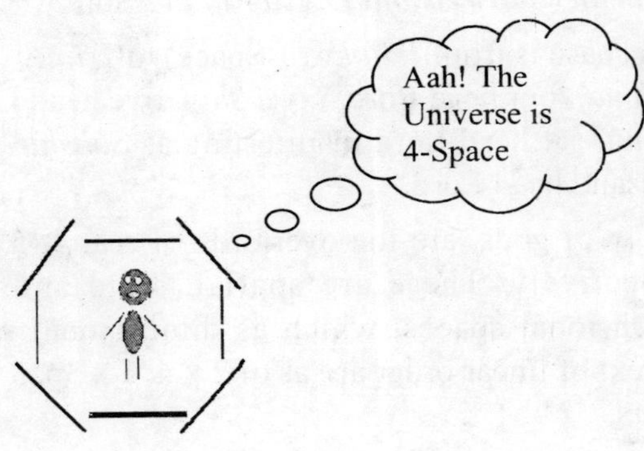

With mind transcending this state and the conditioning manifestation melt, and give way to the emergence of the transcendental state, one gets fulfilled with ambrosia of bliss of the transcendental worlds (5-Space).

This, this way, it is the sequential melting of the mental blocks conditioning the universe for the individuals intellectual ranges which is to be attended to and rest is to followed of its own.

The following, initial information may help the Sadhkas/ Students of Mathematics to have prior acquaintance with the basic technical terms and symbols.

1. Central focus is upon the phenomena of our universe of existence as a space book.
2. *Vedas* are the transcendental fountainhead of ancient wisdom about this space book, as the scripture of 26 basic elements/ *Tatavs of Vishnu Lok*.
3. *Sankhya darshan*, in terms of 25 *tatavs* approaches 26th Tatav as *Atman* (Self)/*Vishnu*/*Divya Pursha*/Sun.
4. This chase is from *Triloky* (3-Space) to *Trinity* of gods (*Lord Brahma*, four head Lord, *Lord Shiv*, five head Lord and *Lord Vishnu*, six head Lord manifesting as *Sahshtra shirsha*, the thousand head Lord.
5. *Trinity* of gods, are the overlords of real 4, 5 and 6 Space respectively. These are spatial, solid and hyper solid dimensional spaces, which as dimensional ranges in the context of linear order are as of 2 x 4, 1 x 3 x 5 and 2 x 4 x 6 orders.

6. *Lord Brahma*, four head lord, is creator the supreme and is the overlord of 4-Space, *Lord Shiva*, five head lord with three eyes in each of His heads, is the overlord of the transcendental worlds of real 5-Space.

7. Creations of *Lord Brahma* are four fold manifestations of four consecutive dimensional space in the roles of dimension boundary, domain and origin. *Lord Shiva* accepts five fold transcendental range with fifth fold as the transcendental fold emerging inward the origin.

8. Within 4-Space, the twelve edged cube acquires additional (13th edge) and with it 4-Space manifests format of the order 13 x 4 = 52 letters alphabets for organization of the knowledge as *Vedas* as creator's manifestations with its potentialities to be of the transcendental order for the transcending mind glimpsing the transcendental worlds.

9. These potentialities of this organization of unfolding transcendence values of knowledge is there because of the letters of *Vedic* alphabet accepting transcendental values as syllables of spatial order with transcendental base. The spatial order along the transcendental base works out as a pairing processing process of Alphabets which being 52 in number organize as 26 basic *Tatav*/syllables.

10. The pairing processing process of spatial order of creator's Space as 26 basic elements/*Tatav*/syllables/artifices of numbers 1 to 26, this way emerges to be the basic *Vedic* mathematics science and technology operation and process being availed by the *Vedic* systems including that of *Ganita Sutras* as well as the core systems of ancient wisdom, as a common pool, with *Vedas* as the core.

11. It may be taken as that it may be for historic reasons that these *Sutras* and systems as well as the processing process somehow had gone out of focus and as such all this went dormant and fortunately with the pointed attention to it and efforts of *Swami Barti Krishna Tirth Ji Maharaj* and also of *Shri Shri Pad Baba Ji Maharaj* and *H.H. Maharishi Mahesh Yogi Maharaj* the process is awakening and is becoming lively.

12. *Sadhkas* with intensified urge to know more and to acquire perfection of knowledge about this process for its pure and applied values shall pair the artifices of numbers 1 to 26 with English Alphabet letters A to Z in that sequence and order and reach at the space book values and get fulfilled with ambrosia of bliss of the transcendental world.

Further the acquaintance with the following arrangement of numbers 01 to 99 as nine columns and eleven rows to help in the chase:

UPPER PART

01	02	03	04	05	06	07	08	09	
10	11	12	13	14	15	16	17	18	
19	20	21	22	23	24	25	26	27	
28	29	30	31	32	33	34	35	36	
37	38	39	40	41	42	43	44	45	
46	47	48	49	50	51	52	53	54	
55	56	57	58	59	60	61	62	63	
64	65	66	67	68	69	70	71	72	
73	74	75	76	77	78	79	80	81	
82	83	84	85	86	87	88	89	90	
91	92	93	94	95	96	97	98	99	100

LOWER PART

The above organization is of two parts, the upper part and lower part. Both parts are self-contained as far as the reflection pairs of numbers are concerned. The upper row 01, 02, 03, 04, 05, 06, 07, 08, 09 accepts the numbers along the diagonal 10, 20, 30, 40, 50, 60, 70, 80, 90 respectively as the reflection pair numbers. These and all the numbers between the first row and this diagonal constitute upper part. The others constitute the lower part. The upper part accepts mirror line of self-reflecting number 11, 22, 33, 44. The lower parts accepts the mirrors line constituted by self reflecting number 55, 66, 77, 88, 99. This is the organization of the format of chapter-13 of *Shrimad Durga Sapat Sati*. This is the basic format of pairing of artifices of two digits numbers. NVF (DISCIPLINE) = 100. Well indicates that it is the artifice of hundred which need be disciplined while all previous artifices already stand disciplined within the organization NVF (ULTIMATE) = 101 is the value of artifices of the order of 101 branches of *Yajor Ved*. There are 1000 branches of *Samved Lord Krishan,* incarnation of *Lord Vishnu*, enlightens in *Shrimad Bhagwad Geeta* as that amongst the *Vedas*, He is *Samved*. Further enlightenment is that amongst the Akshras, He is *Akara*, first letter of the alphabet. When one goes through the formulations One, Two, Three and so on uptill nine hundred ninety nine, one sees that the letter "A" having not been availed. It is for the first time that need for letter "A" having arisen only for the formulation "thousand". *Lord Vishnu* is *Sahshtra Shirsha*, Lord of thousand, (and thousands) of heads.

The acquaintance with the following special symbols may also help in the chase:

1-space	*2-space*	*3-space*	*4-space*	*5-space*	*6-space*
—	□			卐	
$\frac{a}{2}$	$\frac{a^2}{4a}$	$\frac{a^3}{4a^2}$	$\frac{a^4}{8a^3}$	$\frac{a^5}{10a^4}$	$\frac{a^6}{12a^5}$

The organization format of space book truth is creator's Space with transcendental world at its origin.

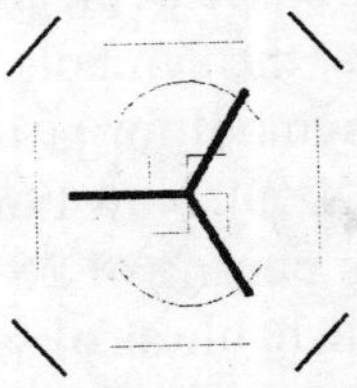

With it, the spatial order of creator's Space stands transformed as solid order. NVF (MONAD) + NVF (MONAD) = NVF (TRIMONAD) is the truth of the transcendental world where NVF (ZERO) = NVF (UNIT) = 64 = 4 x 4 x 4 and the same as trimonad, the artifice 444. This together with the right (solid) units boundary of 4-Space (with domain boundary ratio for hyper cube-4 as $A^4{:}8B^3$) makes the range as 452 units.

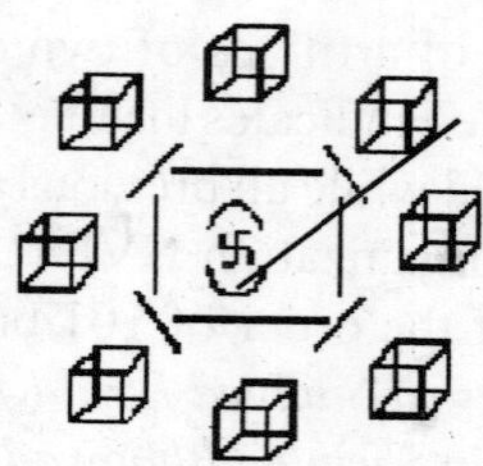

The range of seven colours spectrum flowing through rays of the Sun is precisely of this range:

(1) NVF (VIOLET) = 83 = 73 + 10 = NVF (FORMAT + 10).

(2) NVF (INDIGO) = NVF (TWO) = 58.

(3) NVF (BLUE) = NVF (MIND) = 40.

(4) NCF (GREEN) = NVF (LORD) = 49.

(5) NVF (ORANGE) = NVF (FOUR) = 60.

(6) NVF (ULTRA VIOLET) = 155 = 145 + 10 = 5 X 29 + 5 X 2.

NVF (BLACK) = 29; BLACK PENTAGON = NVF (ORIGIN-ORIGIN).

(7) NVF (RED) = 27 = NVF (HALF). The total range, as such is 83 + 58 + 40 + 49 + 60 + 155 + 27 = 472 = 444 + 28 = 500 - 28.

The everything of this space book is precisely settled; even the forms of the script letters, the symbols of plus (+), minus (-) and so on. Illustratively the symbol for plus (+), as pair of axes. It is NVF (LORD) = NVF (AXES). Now Lords as axes is there as axes are already plural. In this pairing of 26 *Tatav* organization, duals and plurals go in distinguishable as of same generic. NVF (TWO) = NVF (BUNDLE) and NVF (PAIR) = NVF (SPACE).

So NVF (PLUS) = NVF (P) + NVF (L) + NVF (U) + NVF (S) and NVF (S) being for plurals and hence for axes already plurals, makes NVF (PLUS) = NVF (AXES) + NVF (S) and hence the symbol for plus being as it is "+". This chase for plus with its pairing with artifices of numbers as 16, 12, 21, 19 also can be chased but such chase for the formulations is to go out of the range of present stage present study. However, this chase for plus as 16 as 5 + 6 + 5 and then for 12 as 12 boundary components of 6-Space, further 21 as 1 + 2 + 3 + 4 + 5 + 6 and also as 1 x 3 x 7, and still further 19 as ahead of 18, as middle of northern and southern hemispheres, is to complete this chase.

Present study is presented for the Sadhkas having intensity of urge to know the way, knowledge which goes dormant because of manifested organization, can be made transcendental. This chase may help to learn and teach on the universe as manifestation layers quarter by quarter and the layers when chased along artifices of numbers is to begin ahead of artifice of five and to permit the range of nine from the other end of the quarter as transcendental coverage. This is there as quarter of one thousand is to be 250 and the number value formats generics need to be covered are to be 6 to 241. NCF (FIVE) = NVF (NINE) = NVF (NEW) shall be making firstly the range 1 to 250 as of whole quarters then as a recycling process of quarter by quarter is to cover the whole as half and remaining half to be the reflection pair. With this it is only the coverage up till generic of artifice 241 that the whole would stand chased. The bible generic of NVF 241 is NVF (PRESUMPTUOUSLY).

My special thanks are for Kenneth R. Williams for the FOREWORD as well as being the intellectual companion during the chase.

I sincerely feel that the real credit for this pursuit is to go to my all-family members, who have been my source of strength all these days of this pursuit.

The contributions of Shri Lakhdeep Singh, the Computer Operator is self evident, and my all praise goes for his skills and labour.

The appreciation for the nice publication of the volume goes to the publisher.

The study is dedicated to Maharishi Mahesh Yogi Ji Maharaj and whole range of living saints.

12 February, 2004 **Dr. S.K. Kapoor**

श्री ॐ

Learn & Teach Vedic Mathematics

Contents

Section-8

For Cosmic Intelligence Learning from Stage-1 209-248

Section-9

Appendices 249-308

*****

Section I

ANCIENT WISDOM

Sri Om

Learn & Teach Vedic Mathematics

I

Urge To Know

'Urge to know' is a stage and the state prior to the stage and state of learning.

'Urge to know' is 'full eye limit' while 'to learn' is just to 'eye void'. The difference between 'eye limit' and to 'eye void' is there as a difference of 'void-limit'. This is the difference range of 'stage-state' which being because of CENTER and 13-edged cube tagged with it in creator's space (4-Space).

The intensity of 'urge' is a 'state'. The degree of learning is the 'stage'. The intensity of urge is fountained by *Brahman*, the ultimate. The degree of learning is fountained by *Vishnu Lok*. Vedic literature well preserves and enlighten us as that the *Vishnu Lok* is known for its 26 *Tatav*/elements.

Sankhya systems confining themselves to 25 elements of human frame reach at 26th element as *Atman* (soul) as *Vishnu* (God, Sun-God, Creator the Supreme).

Yoga system attain unison by pairing *Atman* and *Brahman* by availing *Ativahkas* (*Brahman* intelligence embedded guided missiles) within rays of the Sun, carrying Being (soul/*Atman*) after its take off from within Sixth *Chakra* (Sixth eternal circuit) of human frame to orb of the Sun, as *Divya Pursha* and then onward to *Sapat Rishi Lok* and thereafter to *Brahman* domain.

This, this way, as first phase, is the phase from Earth to Sun, as of range of 26 elements/artifices of numbers 1 to 26 accepting 'pairing' as unison process which has been fully exploited and worked out by Vedic Mathematics, science and technology and is successful to unify whole range of knowledge from matter to frequencies as a single discipline of knowledge. It is this attainment which in fact is the basis for the claim that whenever human mind would interact with NATURE it would come out with processing which would run parallel to all what has been decipherable in terms of the *Vedic* systems. This grand unification process deserves to be chased.

Credit goes to *Swami Bharti Krishna Tirth Ji Maharaj* to approach *Ganita Sutras* by addressing them the *Vedic* way of reaching at their systems of processing through unison attainments of meditation and the blissful outcome of working rules standing deciphered out in the process are our wonderful treasure. Though unfortunately much has not reached us because of that having been lost but certainly all has not been lost and in fact the core is intact and well preserved in preservation and availability of the processing process. The processing process being that of well preserved Ancient *Yogic* unison processing and the same having been transcendentally approached as *Maharishi Mahesh Yogi's* transcendental meditation technique, and as such, the inner folds of *Ganita Sutras* and other *Vedic* systems approachable in terms of *Ganita Sutra* can be blissfully unfolded by approaching *Vishu Lok Tatav's* along artifices of whole numbers 1 to 26 in terms of the *Ganita Sutra's* systems.

By practicing this technique for approaching first phase from Earth to Sun and by applying the technique of pairing of artifices of numbers, the blissful attainment comes to be as that these artifices of whole numbers 1 to 26 manifest the format for 26 letters Alphabet to reach from SPACE BOOK to SPACE *BIBLE*. The ambrosia of bliss lies in the attainment of revolution as that the orthodox and classical English vocabulary with its every word is a geometrical format formulated form being carried along the artifices of numbers parallel to the frequencies of carriers within the rays of the Sun. And, everything working smoothly as 26 elements, artifices of numbers 1 to 26 and Alphabet letters A to Z in that sequence and order all running parallel to each other and accepting 'pairing' as mathematical operation.

This as A = 1, B = 2, C = 3, ..., X = 24, Y = 25 and Z = 26, shall be working out URGE as U = 21, R = 18, G = 7, E = 5, total 21 + 18 + 7 + 5 = 51, which may be designated and accepted by definition as number value format (in short NVF) of URGE. It may be symbolized and expressed as NVF (URGE) = 51. Then NVF (FULL) = 51.

The 'URGE TO KNOW' with NVFs as 51, 25, 63 and 'FULL EYE LIMIT' as well of NVFs being of 51, 25, 63 may help us comprehend the underline common formats for 'URGE TO KNOW' and 'FULL EYE LIMIT'. Like that whole range of orthodox English Vocabulary can be approached for their formats, with geometric, arithmetic and whole range of technological pure and applied formulations and values.

Sadhkas,

having

(a) an intensified urge,
(b) to know,
(c) and chase,
- the transcendental glimpse of *Vedic* mathematics

shall,

(a) sit comfortably and,
(b) permit the mind to transcend,
- to glimpse the transcendental world,
- as transcendental phenomena,

and have,

(a) self validation of,
(b) the way and path,
- of transcendental glimpse of *Vedic* mathematics.

Between two sittings of trans,

the *Sadhkas,* shall,

(a) go for the intellectual exercise of to be through the scriptures,
(b) and also share the experiential bliss of transcendental world with fellow *Sadhkas,*

(c) and have self validation of experiences, as well as to be sure about the perfection of intelligence gained through transcendence and through intellectual exercise.

Sadhkas for to be sure about one's perfection of intelligence about the discipline of internal folds of the transcendental world shall enlist one's experiences and share them with fellow *Sadhkas* in the sequence those are gained and also for their validation from the scriptures. This exercise, oral or written, would be of great help, at least, at the initial stages, as the transcendence to the inner folds of the transcendental world is a very delicate exercise for the transcending mind and the transcendental world also unfolds and folds back of its own and for it the chase of the transcending mind of the transcendental phenomena of folding and unfolding of the inner folds of the transcendental world and the transcendence progress of the transcending mind, both are to be chased by the transcending mind itself.

The whole range of the scriptures is unfolding as the transcendental glimpse of *Vedic Mathematics* and it is available with the *Sadhkas* to have it as transcendence range for satisfying the urge and fulfilling the transcending mind with the ambrosia of bliss of the inner most fold of the transcendental world.

II
Learn and Teach

NVFs range SPACE BOOK to SPACE *BIBLE* is 44 - 43 to 44 - 30. The NVFs difference of SPACE and BOOK as 44 - 43 = 1 and of SPACE and *BIBLE* 44 - 30 = 14 is the range of *Ganita Sutra*-1 and *Ganita Sutra-14*.

NVF (LEARN) = 12 + 5 + 1+ 18 + 14 = 50 = NVF (VOID). The de-void-ing process is of taking out unit by unit. It is to be of the rule, sequence and order of 'One less than before' as is the processing system of *Ganita Sutra*-14 '*Ekanunena Purvena*/One less than before'. NVF (DE) = 4 + 5 = 9 is *Brahman*. Further NVF (NINE) = NVF (NEW). Then, NVF (ING)= NVF (*BIBLE*).

NVF (DEVOID) = 59 = NVF (SOLID) = NVF (LINEAR). It is to take out solids, as a linear process, one at a time.

Modern systems have yet to address itself as to what is the dimension of 1-Space. Up till this time, modern systems are approaching 1-Space simply as one axis of three-dimensional frame. It is yet to comprehend by the modern systems as that (—1) Space is playing the role of dimension of 1-Space. It is like that is written SPACE BOOK; NVF (SPACE) = 44 and NVF (BOOK) = 43 and it is the order of measures to go one step down as that the tree of world is with its roots upward and stem increasing downward as is the enlightenment preserved in *Shrimad Bhagwad Geeta*.

To learn means 'Eye void' as NVF (TO) = 35 = NVF (EYE) and NVF (LEARN) = 50 = NVF (VOID). The void, which is to be further de-voided, is parallel to *Avyakta* /un-manifest and *Ayakto-Avyaktat* /base of un-manifest as is the enlightenment of *Shrimad Bhagwad Geeta*. This takes to the boundary of transcendental world/ 5-Space/ hyper cube-4 which accepts domain boundary ratio as A^5:$10B^4$ and requires 10 x 5 = 50 coordinates to fix transcendental world within its boundary as void/ un-manifest as comparison to cube/solids as manifested bodies. NVF (EYE) = 35 and this artifice as five at unit place and three at next place is 5-Space with 3-Space as dimension and further its re-organization as 5 x 7 parallel to seven hyper cubes-5 manifesting dimensional frame for 7-Space as unifying format of unity state of consciousness makes it out how VOID is to be to FACE EYE as both being of same generic NVF with NVF (VOID) = 50 = 15 + 35 = NVF (FACE EYE) and here FACE as of NVF 15 of the order 1 x 3 x 5 of dimensional world (5-Space of solid order that is 3-Space in the role of dimension and 1-Space in the role of dimension of dimension).

This as such focuses as to, what is to be learnt, as void being transcendental world itself de-voided of all manifestations from within the space/*Akash*/ Fifth *Mahabhhut*/ 5-Space.

To teach would mean to 'EYE CONE' as NVF (TEACH) = 37 = NVF (CONE).

Learning focus as VOID of FACE -EYE range and teaching focus as CONE (with FACE) in fact settles the learning - teaching range as Earth to Sun. As NVF (EARTH) = 52 = 15 + 37 and it is two units more than NVF (VOID) as a middle gap of two points in the NVF measure range of Earth. Likewise NVF (SUN) = 54 makes it to be a two points more than NVF (EARTH) as the same being a middle gap of two points in the NVF measure range of Sun.

Learning face to EYE VOID and teaching face to EYE CONE become the key reminders for the *Sadhkas*/Students of *Vedic* Mathematics having intensity of urge to know and to learn for teaching.

III
Four Courses

Initial stage study material has been subject matter of four free VM Courses being taken up through web-site (www.learn-and-teach-vedic-mathematics.com).

Vedic Mathematics Course-1 is titled "Vedic Mathematics On Geometric Formats of Real Spaces". Lesson 1 to 126 are uploaded on above website and the same may be downloaded. List of the lessons is enclosed as Appendix-I of this volume.

Here below is reproduced lesson-1 of this course to give an idea of the nature of the subject content:

Learn And Teach Vedic Mathematics

VM COURSE-1

Lesson 1 - Float "SPACE thought" for the TRANSCENDING mind.

Section-1 Lesson Text

Section-2 Initial leads under this Lesson

Section-3 Original documents file

Section-4 Announcement

Section-1 Lesson Text

Learn and Teach Vedic Mathematics on geometric formats of real spaces of Vedic comprehensions preserved in Vedic scriptures

---

Lesson 1

[as on 15 August 2003]

Float "*SPACE* thought" for the *TRANSCENDING* mind.

Background:

Akash (अकाश:), Nad (नाद:), Jyoti (ज्योति:) and like are all designations of Braham and as such different approaches to Braham.

Lesson steps

1. Sit
2. *(i) comfortably,
3. Permit the mind to transcend *(ii)
4. Float "Space thought" for permitting the transcending mind to avail these frequencies of AKASH (SPACE as THOUGHT) as carriers to the transcendental base of all manifestations as, as well as within SPACE. During transcendence, if one becomes conscious as that the transcending mind has missed to be carried by the frequencies of Space thought, then one can remind oneself again (and again) by silently reciting this thought for the transcending mind to avail space thought frequencies to transcend further and further to reach and glimpse the ultimate to which space thought frequencies carry. This is very simple for everyone to remind oneself like that. However one can take help of senior Sadhkas. The fortunate one's may go to the teachers and learn it. One's faith and practice puts everything on the track of its own.
5. Once the transcending mind glimpses the ultimate and is full-filled with the bliss of the ultimate and ascends, the bliss is to be enjoyed and shared with other Sadhkas as well as the experience to be self-validated in terms of the experiences of other Sadhkas well preserved in the Scriptures.
6. One should continue compiling the experiences of other Sadhkas preserved in the scriptures*(iii) about Akash till one

again has an urge for further transcendence for fresh glimpsing of the ultimate availing frequencies of Akash (space thought) as carriers for the transcending minds and this to continue again and again till full satisfaction with ambrosia of bliss completely fulfilling the transcending mind to its full to the brim.

*

(i) One may take help from Shalokas 11 to 13 of Chapter-6 Shrimad Bhagwad Geeta.

(ii) One may take help from methodologies of transcendence through meditation, particularly the transcendental meditation enlightened by Maharishi Mahesh Yogi and other great Rishi's enlightening about Vedic meditation methodologies.

(iii) Swateshwatra Upanishad in its canto 2.11 onwards preserves experiential truths of Sadhkas about initial stages through which the transcending mind would pass through and the subsequent attainments. The Sadhkas may avail this and other references for perfection of intelligence through experiential truths.

Section-2

Initial LEADS under this Lesson for further interests in the topic

Background

Akash (अकाशः), Nad (नादः), Jyoti (ज्योतिः) and like are all designations of Braham and as such different approaches to Braham.

A word for Sadhkas

This course as the title suggest is for Learning and Teaching Vedic Mathematics on Geometric Formats of Real Spaces of Vedic comprehensions preserved in the Vedic scriptures. As such the methodology of approach is the Vedic way of experiential comprehensions of truth of reality in space of our existence.

With this Akash (Space) merges our beginning as well as the end. Therefore the first lesson, ultimately is going to be all spreading throughout. This being so the urge of the transcending mind to glimpse the ultimate of Akash (Space) be permitted to be satisfied by allowing the mind to transcend as many times and at any time of such urge throughout the Course of one year with which would be completed one another revolution of the Earth around the Sun and with it there would be further churning of universal bliss.

Introductory

Here Akash as designation of Braham has been accepted for approach to Braham.

Akash as Braham, Akash as Akashtatav and Akash as fifth Mahabhut are the different features of this designation of Braham.

The Akash (Space) as Brahmand; as sustaining domain of cosmic egg for whole range of creations and Vishwa (world sustaining existence) is the Reality feature of existence, which makes Akash as a Real space. Akash as real space with all Panch Mahabhut (Solids, liquids, fires, Airs and Spaces) as different five folds of Akash tatav (Space element) individually emerge as five distinct Real spaces.

Lesson focus

As every system deserves approach acceptable to the system, so the Vedic systems deserve to be approached in Vedic way only, and as such the methodology of the approach being presumed is the Dhayan-yog-anugatah (ध्यानयोगानुगतः): by being established in Yogic meditation state and the preliminaries of posture etc. being a comfortable sitting in Yogic posture with body (head, neck, chest, straight high) senses established by mind in the heart and availing the format of Om as Brahman boat for carrying through the worldly oceans.

Frequencies of Akash

The formulation अकाशः (Akash)= अक्+आश्अः (Ak+ash) and अक् (Ak) as प्रत्याहारः (Pratyahara) beginning with first vowel अ

(a) and ending with first consonant क् (k) takes us through first and second Maheshwara Sutras (अइउण् and ऋलृक्) and thereby the coverage range extends up till first five vowels (i) अ (ii) इ (iii) उ (iv) ऋ (v) लृ. Therefore Akash (अकाशः) Space of this comprehension and visualization is reality which is chaseable as limit of five basic Swaras (स्वरः त्र स्+वरः, that sustains itself) / vowels.

*****

IV
Vedic Mathematics for Beginners

Vedic Mathematics Course-2 is titled "Vedic Mathematics for Beginners. Lesson/Aspects 1 to 112 are uploaded on above website and the same may be downloaded. List of the lessons is enclosed as Appendix-II of this volume.

Here below is reproduced lesson-1 of this course to give an idea of the nature of the subject content:

Aspect 001- ABOUT CONSTRUCTION OF EKADHIKENA NUMERALS

Vedic Mathematics for Beginners (as conceptual skills approach)

About the Course

1. The Learn and Teach Vedic Mathematics on Geometric Formats course is in its third week. The aim is to settle the frontiers of mathematics of Vedic comprehensions. In a way this is going to be an attempt to reach at the final shape of things.
2. However, in the present conceptual skills course of Vedic Mathematics for beginners, the aim is to approach from the very beginning parallel the growth of intellect of the child.
3. Here in this Vedic Mathematics course for beginners, the fundamental theme is (that)
   (i) Child is the latest arrival in the family from abode of God.
   (ii) The grace on the face, virtue in the eyes and innocence all around the child is the gift of the God put under care of the parents (including all those who accept the duty as responsibility for growth) as solemn trust.
   (iii) The duty and responsibility of teaching for growth of the child, as such, as well, comes well within the solemnity of trust.

4. This being so, the present teaching course through conceptual skills approach of Vedic Mathematics, with all seriousness, is to come up to the expectations of the sensitivity of solemnity of the trust.

   And, further, this conceptual skill approach is also to be as natural and sublime, virtuous and pure, as well as solemn and truthful taking enough care that neither the grace gets coloured, nor virtue gets diluted nor innocence gets screeched in any way during teaching with this approach of conceptual skills. For it the 'teaching with kindness and learning with love' are the key words for the teachers and students of Vedic Mathematics.

5. One of the basic tenets of this approach of teaching and learning through conceptual skills is to progress step by step, with one step at a time, in continuity of the previous step in a way to give way for smooth transition for the following step and thereby attaining complete system for continuity for the whole range from beginning to the end of learning.

6. Further, it is acknowledged as that the accepted ideal is to comprehend and chase the concepts in their purest forms. However the practical approach of beginning from the very beginning of formal education of child demands that the beginning is to be had in terms of the skills as applied values of the pure concepts.

7. As such here, the organization format of Ganita Sutras, is accepted as reference frame for the chase. The different formats adaptations for the skills during different phases of time as well may be touched at appropriate stages of progress of this course (as conceptual skills approach).

8. Naturally the ultimate focus is to be the conceptual formats of Vedic Mathematics. But, to them, in this approach, learners can revert only after the firm grip over the skills is achieved. As such, in this approach, there would be a need of extra care to see that the process of learning the skill to see that this in no way comes in the way of ultimate comprehension of the concepts in their purest forms.

9. With all earnestness course of this approach is being launched with hope of serving the cause of the Vedic Mathematics.
10. The pooling of experience of senior teachers, scholars and textbook writers is solicited to contribute a set of lessons to cover a specific topic as guest expert.
11. This Course is being designated as for beginners simply for the reason that here aim is to begin from the very beginning. Otherwise, the range of the course is optimistic from Arithmetic to Astronomy and it is to complement and supplement the main course of learning and teaching Vedic Mathematics on geometric formats. As such it is to be approached with a respect of a foundation course whose idea is to prepare a ground for chasing frequencies within the rays of the Sun along the formats of anshubodhni (अंशुबोधिनि) of Rishi Bhardwaj (ऋषि भारद्वाज).
12. Vedic Mathematics, by definition, be taken as Mathematics of Vedic systems, which include Ganita, Sankhya, Sathpatya and further the whole range of systems of Vedic knowledge.

First Week of the Course
(29 August 2003 to 04 September 2003)
(Aspects 001 to 007)

Topic of the Week

Success of the course depends upon the kindness with which teachers teach, love with which students learn and consciousness of parents with which they chase the intellectual growth of their children. As such success foundation prayer is:

Success foundation prayer

Teachers	Students	Parents
Be kind Teach with kindness, Step by step, and each step to be	With love Learn with love, Step by step, learning each phase of every step as per	Be fully conscious Have conscious growth chase, Step by step, ensuring expected

phased for proper leads; and each lead being a smooth take off, for transcendental domains

the leads; to have smooth take off for transcendental domains

intellectual growth and perfected intelligence of transcendental domains.

Aspect 001

ABOUT CONSTRUCTION OF EKADHIKENA NUMERALS

Source: Ganita Sutra-1

Concept: One more than before

Format: 1 as 0 + 1
2 as 1 + 1, 3 = 2 + 1, 4 = 3 + 1 and so on

Skill: To construct Ekadhikena numerals

Part 1
General

A WORD WITH PARENTS

Your child is the latest arrival from the abode of God. The grace on his face, virtue in his eyes and innocence all round is godly trust for which it is your responsibility to remain fully conscious throughout the education of your child and have conscious growth chase of your child. The informal but firm education pre-school growth of your child is your responsibility, which extends teaching and ensuring proper comprehension of two fundamental concepts of association of one with one thing and zero to be as no-thing, (nothing). The symbols 1 and 0, as line and circle (or a close curve in plane) as well as the names "one" and "zero" for them as well are included within the domain of responsibility. With this education, the entrustment of child in the hands of enlightened teacher full of kindness, is to ensure proper mathematical education for your child. As such it is solicited that parents shall come up to the godly responsibility.

A WORD WITH THE TEACHERS

Your enlightenment and kindness has bestowed responsibility upon you for initiation of formal mathematics education of innocent children requiring growth and perfection of intelligence under your instruction. You have to make them believed 0+0=0, 0+1=1, 0x1=0, 1x1=1 and 1+1=2. Further you have to smoothly expose to the technique and skill of reaching up till the tables 5x5. Then you have further the responsibility to teach the skill of reading the symmetry of the formations of the body of the child as the FIRST BOOK OF MATHEMATICS brought by the child along with him from abode of God. This book is to be taught to be learnt with the help of a looking mirror for perfection of the skill of observation (विलोकनम् Vilokanam). It is solicited that these three initial responsibilities be attended to with all softness of kindness for pleasant comprehension by the children with love.

A WORD WITH THE STUDENTS

God has chosen Parents for you. Your Parents has chosen Mathematics teacher for you. You be good enough to respect them with your love for all kindness and attention of parents and teacher for your proper growth. Accept with love the kindness of teaching and the skills to be taught to you. It is your love for your teacher, which is to ensure flow of kindness for you from the core of the heart of your teacher.

Part 2

Introduction of the skill to be learnt

(i) This skill (conceptual skill aspect 01; in short VM Skill-01) is about construction of Ekadhikena numerals.

(ii) Ekadhikena, by definition, be taken as "one more count".

(iii) Numerals are (1, 2, 3, 4, 5, 6, 7, 8, 9).

(iv) By definition, we may accept 10 as also numerals only for the purpose of construction of Ekadhikena numerals for the numerals 1, 2, 3, 4, 5, 6, 7, 8, 9.

(v) Ekadhikena numeral of a given numeral, by definition

means a numeral of one more count than the count of a given numeral.

Part 3

Source concept

(i) Source Sutra is Ganita Sutra-1 Ekadhikena Purvena (एकाधिकेन पूर्वेण). The simple English rendering for its working rule is: (By) one more than before.

(ii) The concept beneath the working rule is the concept of ordering with given value as a count. The working rule comes to be to have increase of values as counts, by count step following a count step.

(iii) The numerals 1, 2, 3, 4, 5, 6, 7, 8, 9 are nothing but the counts, manifesting in terms of the rule and concept of Ganita Sutra-1.

(iv) The numerals, as counts (and as values), being manifestations of count steps following a count step, as such Vedic systems, also chase the phases of the working steps in terms of which the numerals are reached at. The outcome of this phasing are the Ekadhikena numerals.

Ekadhikena Numerals

The first ordering rule of Ganita Sutra-1 as 'One more than before' has been so fundamental that for the purposes of skills to approach mentally, Arithmetic operations, the Vedic system, at first instance as enlarged the family of numerals by including Ekadhikena numerals.

DEFINITION

By definition, corresponding to each numeral, there exist Ekadhikena numeral, as being of value one more than the given numeral.

ILLUSTRATIONS

Illustratively ekadhikena numeral corresponding to numeral-1 is 1+1=2. Likewise Ekadhikena numeral corresponding to numeral 2 is 3; corresponding to numeral 3, it is 4 and so on.

SPECIAL SYMBOLS

The Ekadhikena numerals have been allocated special symbols. These symbols are the symbols of the numerals but with the additional feature of having dots at the tops of the symbols, as under:

Numeral	Ekadhikena numeral	Symbol for numeral	Symbol for Ekadhikena numeral
One	One more than one	1	$\dot{1}$
Two	One more than two	2	$\dot{2}$
Three	One more than three	3	$\dot{3}$
Four	One more than four	4	$\dot{4}$
Five	One more than five	5	$\dot{5}$
Six	One more than six	6	$\dot{6}$
Seven	One more than Seven	7	$\dot{7}$
Eight	One more than eight	8	$\dot{8}$
Nine	One more than nine	9	$\dot{9}$

EKADHIKENA OF PLACE VALUE DIGIT

By definition Ekadhikena of 0 is of value 0+1. The special symbol for Ekadhikena of zero is taken as: 0a

Accordingly, with inclusion of Ekadhikena numerals, the family of numerals gets enlarged as of double number of members. This enlarged family has following twenty members:

Numerals	1	2	3	4	5	6	7	8	9	0
Ekadhikena Numerals	$\dot{1}$	$\dot{2}$	$\dot{3}$	$\dot{4}$	$\dot{5}$	$\dot{6}$	$\dot{7}$	$\dot{8}$	$\dot{9}$	$\dot{0}$

Note:

The Ekadhikena of nine and of zero deserve special attention. As nine is the largest numeral of ten place value system, as such Ekadhikena corresponding to nine is to be ten, which is the place value, and as such is of two-digit expression. Which distinguishes it from all other Ekadhikena numerals remaining as of single digit values. The Ekadhikena of place value digit (0) is of special features as Ekadhikena accepts value of first numeral by itself accepting placement value (as value before for the ordering rule of Ganita Sutra-1).

For operations with the help of Ekadhikena numerals, we have to wait till the further enlargements of the family as accepted by Vedic systems stands introduced.

V
Mathematics Chase of Sanskrit

Vedic Mathematics Course-3 is titled "Mathematics Chase of Sanskrit". Lesson/Chase steps 1 to 84 are uploaded on above website and the same may be downloaded. List of the lessons is enclosed as Appendix-III of this volume.

Here below is reproduced lesson-1 of this course to give an idea of the nature of the subject content:

Chase Step 001- Forty two letters coordinated by 14 Maheshwara Sutras

Mathematical chase of Sanskrit language
(Mathematics of Sanskrit)

About the Course:

1. The present course: Mathematical chase of Sanskrit language is with an aim to reach at different mathematical formats for different aspects of Sanskrit language collectively as a Discipline of Mathematics of Sanskrit language.
2. It is hoped that by reaching at the Mathematics of Sanskrit language, it may be possible to have a clear insight about the organization of Vedic knowledge, its systems, and applied values.
3. One of the reasons that Vedic fort is still invincible for the modern mind is that the Mathematics of Sanskrit has not been appropriately approached and as such this course may help in this direction.
4. Amongst the different approaches chase to Vedic knowledge, the grammar approach is the most fundamental approach for the knowledge organized as Samhitas and following literature as written language.
5. This approach to Vedic knowledge in terms of Grammar, in its widest range covers all aspects of written language and in a way ultimately to cover the entire range of auxiliary

branches/Upangas, (Shiksha, Grammar, Nirukat, Chandas, Jyotish, Kalpa) including the very manifestation of alphabet.

First Week of the Course
(26 September to 02 October 2003)
(Chase Steps 001 to 007)

Topic of the Week: About Alphabet

About the Alphabet:

6. The alphabet is the starting point as well as the end point of Vedic systems. Shrutis to Alphabet and Alphabet back to the Kalpas and ultimate source is the two fold approach to the Vedic knowledge.

7. As such here, the starting point is to be the Alphabet and it is aimed to sequentially reach up till the Kalpa-Upanga.

8. About Alphabet as well there are many aspects like script, pronounced frequencies packages, formats of colours as light frequencies packages, geometric formats for meanings etc. etc.

9. Even regarding the number of letters constituting Vedic Alphabet, as well there are different preservations and as such these are to be chased in terms of inner evidence, as the actual alphabet letters availed by different Vedic Samhitas and other Vedic literature including the books of Grammar, Chandas (meters) etc. etc.

10. Upanishads, in particular Akshmaliko Upanishad and Puranas, illustratively Sakand Puran and Shrimad Bhagwad Mahapuran, Shiksha Upanga etc. specifically deal with Vedic Alphabet in its different aspect.

11. Shrimad Bhagwad Geeta enlightens as that "Lord Krishna commands: amongst Akshras, I am Akara"

12. The first consonant "Kakara" is Chaturmukhi/of four mouths (creator Lord Brahma). Also the meanings of this consonant are: Panchmukhi/ of five mouths (Lord Shiv). This in a way is the initial indication as that the same alphabet as a system can be applied within different dimensional spaces for the

language of Mathematics, science and technology of concerned spaces.

Text of Chase Step-001
Forty two letters coordinated by 14 Maheshwara Sutras

Chase Step 001

Forty two letters coordinated by 14 Maheshwara Sutras

General:

There are fourteen Maheshwara Sutras. Their text (in Devnagri Script) is as:

Sr. No.	Maheshwara Sutras
1	अ इ उ ण्
2	ऋ लृ क्
3	ए ओ ङ्.
4	ऐ औ च्
5	ह य व र ट
6	ल ण्
7	ञ म ङ़ ण न म्
8	झ भ ञ
9	घ ढ़ ध भा्
10	ज ब ग ड द श्
11	ख फ छ ठ थ च ट त व्
12	क प य
13	श भा स र
14	ह लू

Forty-two Devnagri letters coordinated by Maheshwara Sutras:

1. Devnagri Alphabet (of Sanskrit language) has letters other than forty-two letters coordinated by Maheshwara Sutras as well, however here first of all are being introduced only these forty-two letters.
2. These forty-two letters accept two fold classification, firstly as vowels and secondly as consonants.
3. There are nine vowels of Devnagri Alphabet which are coordinating in terms of first four Maheshwara Sutras, viz. in column 3:

Sr. No.	Maheshwara Sutras	Vowels Coordinated in the Sutra
1	अ इ उ ण्	(i) अ (ii) इ (iii) उ
2	ऋ लृ क्	(iv) ऋ (v) लृ
3	ए ओ ङ्.	(vi) ए (vii) ओ
4	ऐ औ च्	(viii) ऐ (ix) औ

4. There are thirty-three consonants of Devnagri Alphabet which are coordinated in terms of fifth to fourteenth Maheshwara Sutras, viz. in Column 3 below:

Sr. No.	Maheshwara Sutras	Consonants in the Sutras
5	ह य व र ट	(i) ह (ii) य (iii) व (iv) र
6	ल ण्	(v) ल
7	ञ म ङ ण न म्	(vi) ञ (vii) म (viii) ङ (ix) ण (x) न
8	झ भ ञ	(xi) झ (xii) भ
9	घ ढ़ ध भ्	(xiii) घ (xiv) ढ़ (xv) ध
10	ज ब ग ड द श्	(xvi) ज (xvii) ब (xviii) ग (xix) ड (xx) द

11	ख फ छ ठ ध च ट त व्	(xxi) ख (xxii) फ (xxiii) छ (xxiv) ठ (xxv) ध (xxvi) च (xxvii) ट (xxviii) त
12	क प य	(xxix) क (xxx) [illegible]
13	श भा स र	(xxxi) श (xxxii) भा (xxxiii) स
14	ह ल्	ह (this letter is the first consonant above)

5. These thirty-three consonants are further classified in two parts, part-1 as 25 Varga consonants and 8 other consonants.
6. The twenty-five Varga consonants accept placement in the alphabet as 5x5 rows column format and that is why these are designated as Varga (square) consonant as accepting 5x5 square format as:

Sr. No.	Consonants	Designation
1	क् ख् ग् घ् ङ्	क-वर्ग (Ka-varga)
2	च् छ् ज् झ् ञ्	च-वर्ग (Cha-varga)
3	ट् ठ् ड् ढ् ण्	ट-वर्ग (Tta-varga)
4	त् थ् द् ध् न्	त-वर्ग (Ta-varga)
5	प् फ् ब् भ् म्	प-वर्ग (Pa-varga)

7. The non-varga consonants also get divided into two parts, part-1 Anthastha consonants and part-2 Ushmana consonants viz.

Sr. No.		Non verga consonants
1	Anthastha	य् र् ल् व्
2.	Ushmana	श् भ् स् ह्

Special Note:

1. The other letters (other than above 42 letters) to be taken up afterwards.

2. The first exercise, naturally is to have script acquaintance with the letters.
3. The script acquaintance is to be complete.
4. One should be able to write down the letters in script form.
5. The second exercise is to know how to pronounce these sounds (packages) as these letters are.
6. This learning to pronounce is a specific step towards education of Sanskrit language.
7. This discipline and science of pronunciation of Sanskrit Alphabet letters is designated as SHIKSHA.
8. There are more than one ancient texts/books of this discipline.
9. Here at the stage before taking up the above two exercises of script writing and sounds pronouncements, the formal acquaintance with the above letters and their broad classification and designations as vowel/consonant, verga consonant/ non-verga consonant, Anthstha/Ushmana etc. to be attempted.
10. From next chase step, further chase of the placements of the letters in the Alphabet format and their classification as sound packages for pronunciation is to be taken up.

Further Study

Lessons 036 to 042 of Course-1 may provide initial insight into Vedic knowledge systems. Likewise aspects 029-035 of course second due this week may further help to have initial insight about the beginners approach to formal education of Vedic Mathematics systems. It may be advisable to go through lessons/ Aspect of other Courses as well as all the four courses are complementary and supplementary of each other. The same (lessons and aspects of Courses 1 & 2 may be downloaded). Here is reproduced Lesson 036 for ready reference study.

VI
Transcendental Basis of Human Frame

Vedic Mathematics Course-4 is titled "Transcendental basis of Human Frame". Lesson/Flow progress stages 1 to 84 are uploaded on above website and the same may be downloaded. List of the lessons is enclosed as Appendix-IV of this volume.

Here below is reproduced lesson-1 of this course to give an idea of the nature of the subject content:

Flow Progress Stage 001- Sathpatya Measuring Rod

Transcendental basis of Human Frame
(Mathematics of Self-referral sustenance)

About the Course:

1. The present course: Transcendental basis of human frame is the discipline which aims to reach at the self-referral sustenance thereof.

2. This course as such is meant for having a clear insight about the way the transcendental basis of human frame is sustained as self-referral order with orb of the Sun as the source.

3. The source state of the self-referral sustenance of transcendental basis of human frame is designated as "Divya Pursha" state within orb of the Sun.

4. Amongst the different approaches chase to Vedic knowledge, the sathapatya approach is the most fundamental approach for the knowledge organized as Up-Vedas and the following literature of self-referral sustenance of existence within frames.

5. This approach to Vedic knowledge in terms of Sathapatya, in its widest range covers all aspects of Up-Vedas of all the four Vedas and in a way ultimately to cover the entire range of Up-Vedas range of knowledge: Sathapatya Up-Ved, Gandharav Up-Ved, Dhanur Up-Ved and Ayur Up-Ved.

First Week of the Course
(26 September to 02 October 2003)
(Flow progress stages 001 to 007)

Topic of the Week: Sathapatya (manifestation format)

About the life flow on manifestation formats:

1. "Sthan (स्थान)" means place; place is space; specific space; space.
2. "Sathapatya (स्थापत्य)" means manifestation on a format; and as such, may be accepted as "manifestation format".
3. As such here, the starting point is to be the formats for manifestations, illustratively as bodies, specifically the human body, with form and frame of distinguishable feature from forms and frames of other bodies/manifestation accepting different formats.
4. About manifestation formats in general and even specifically about manifestation format of human body, as well there are many aspects for whose chase as an existence (as life flow range) is the underline urge for whose satisfaction is going to be the chase (as life flow progress stages), within human frame.
5. The Sun as the source for self-referral sustenance of the transcendental basis of human frame of Vedic comprehensions, as such becomes the source and goal of this pursuit.
6. As such the attainment goal of the Sathapatya approach to Vedic knowledge gets settled as the starting point and the orb of the Sun as the source and goal as the end point of this pursuit.
7. The Sathapatya approach to Vedic knowledge for coverage of this range of life flow on manifestation format within human frame begins with the settlement of the measuring rod with Lord Vishnu as presiding deity of the measuring rod and Lord Brahma as the presiding deity of the measure of this measuring rod.

Text of Flow Progress Stage-001
Sathapatya Measuring Rod

General:

1. Mansara (Man means measure and Sara means essence; the essence of measure) is the scripture of Sathapatya order.
2. True to the title, Mansara (the essence of measure), the scripture accepts its starting point as prayer to Lord Brahma, the creator the Supreme, and concludes with attainment of chiseling of third eye, and the consequential transcendental transition from creator's space (4-Space, with Lord Brahma as the presiding deity) to transcendental space (5-Space, with Lord Shiv as the presiding deity).
3. This transcendental phenomena of transition from spatial order of creations (including human frame) to solid order for transcendental basis for the creations (including human frame) as such becomes the basic object and aim of the Sathapatya order.
4. The Sathapatya order as an essence of measure further makes it imply clear about its starting point as a measuring rod with Lord Brahma, the creator the Supreme and presiding deity of 4-Space, as the presiding deity of measure and Lord Vishnu, the lord of self-referral sustenance systems and presiding deity of 6-Space, as the presiding deity of the measuring rod as well as the wood (content) constituting the measuring rod.
5. This as such is the enlightenment as that the measuring rod with Lord Vishnu as presiding deity, is of the order of self-referral sustenance (within 6-Space) of the transcendental basis (5-Space) of human frame (created by creator, Lord Brahma, the presiding deity of 4-Space).

Introduction of measuring rod:

1. The "Sathapatya measuring rod" with Lord Vishnu as its presiding deity and Lord Brahma as presiding deity of measure of this measuring rod, is very rich basic concept which deserves to be comprehended fully to here and insight about life flow within human frame, in particular.

2. Lord Vishnu as presiding deity of 6-Space and Lord Brahma as presiding deity of 4-Space, helps us have comprehensive view of inter-relationship of 6-Space and 4-Space as inter-relationship of domain and dimension as much as that 4-Space plays the role of dimension of 6-Space.

3. The domain being 6-Space as such it shall be accepting measuring rod constituted by representative regular bodies of 1 to 6 Space.

4. In other words representative regular bodies of 1 to 6 Space constitute the measuring rod for domain (of 6-Space) while representative regular bodies of 1 to 4 Space constitute the measuring rod for 4-Space.

5. The dimension to supply the measures is the essence of measures accepted by the measuring rods is very is the first comprehension of Sathapatya approach which deserves to be imbibed well. This comprehension in reference to 3-Space means measure to be supply by linear axes as linear dimensions of 3-Space. In general (n-Space) as dimension of (n+2 Space) is to supply the measure for the measuring rod of (n+2 Space).

6. As such for the self-referral sustenance of transcendental basis of human frame accepts 4-Space as the source of measures and this that way becomes the first basic comprehension about the sustenance of transcendental basis of human frame in terms of which the life flow within human frame is to be measured.

Note: See Appendix I to IV for lists of Lesson of above referred four free VM Courses on website (www.learn-and-teach-vedic-mathematics.com).

Appendix-I List of 126 Lessons of Course-1
Appendix-II List of 112 Lessons of Course-2
Appendix-III List of 84 Lessons of Course-3
Appendix-IV list of 84 Lessons of Course-4

*****

VII
Why Vedic Mathematics?

The questions which should be addressed are as to why the modern mathematics is held up, why its logic recoils upon itself and why there are mathematical problems, logical knots and mental blocks at all in the modern mathematical approach?

Well known problems of modern mathematics may be cited as:

1 Everywhere continuous but nowhere differentiable functions
2 Hypercubes 1 to 7 increase but hypercube 8 onwards decrease
3 Space Filling Curves
4 Riemann Hypothesis
5 Goldbach's Conjecture
6 Fermat's Last Theorem

Isn't it that these problems are there because of the axioms accepted by the modern mathematics?

And then follows a question as to whether Vedic mathematics is in a position to help the modern mathematics to come out of its mental block and to un-tie its logical knots and to solve the problems?

The Vedic geometric concepts worked out in the books of Dr. Kapoor promise us geometric comprehensions of our existence phenomenon transcending our existing three-space format. The real four and higher spaces formats of Vedic comprehensions are new wonderful worlds of very rich mathematics which may ensure us powerful technologies and much potentialised disciplines of knowledge. The basic comprehension pointed out is the way the cosmic surface constitutes and binds the solid granules as synthetic solids manifesting in the cosmos.

Dr. Kapoor is attempting to reconstruct the discipline of geometry as a discipline based on Vedic concepts. He has designated this discipline as Vedic Geometry. His results has added a new dimension to the dialogue initiated with the interpretation of the Ganita Sutras and their potentialities brought to focus by Swami Bharti Krisna Tirthaji Maharaj.

Dr. Kapoor's conclusion is that this all is there only because of the acceptance of the geometric entity (monad) admitting no parts, and "1" has no predecessor. To overcome this, as per him, the modern mathematics needs Vedic mathematics' help to shift from monad without parts to a monad admitting parts. The eliptic equations format $y^2=x^3$ is bound to give a conceptual slip and this, as per him, can be well glimpsed by chasing the format of this equation on simplex format to see how it is deceptive to appear to be so while as whole numbers artifices parallel to the dimensional frames is well evident inequality. As such, there is a need for the modern mathematics to re-address to itself about the need for re-settlement of the basics to come out of the mental blocks and logical knots to un-tie the knots and to transcend the blocks and to be face to face with the wonderful worlds of reality awaiting ahead with all potentialities of their structural richness. The parallelism between artifices of whole numbers 1 to 26 and 26 sporadic groups is there because of the cosmic surface within the solids.

The recent academic research attempts and teaching experiments with the help of Vedic mathematical operations demonstrate their potentialities to provide the desired help.

The research results are bringing us nearer the traditional acceptance as that Vedas are written on the rays of the Sun. Vedic mathematics, science and technology is the mathematics, science and technology of the way the nature maintains grand unification of the existence phenomenon on the Earth through the rays of the Sun. It is in this grand design of the nature the individual Vedic mantras are impulses of consciousness. This design maintains the continuity of the life within human frame and beyond through the natural intelligence embedded in the human mind and in the rays of the Sun. This continuity and parallelism when chased promises new wonderful experiential domains about new realities and the wonderful domains to unfold for us new disciplines of mathematics, science and technology.

Vedic sounds are multidimensional domain frequencies from within the particular dimensional frame as the structure of that domain. When the sounds are pronounced, the frozen frequencies get initiated and the self-organizing power of the Vedic sounds set the frequency's potentialisation process into action. It is this process whose utilization is the aim of different Vedic scriptures.

Rig Ved Samhita is the first Vedic scripture. It is the first book of the mankind. The mathematics precedes the composition of Rig Ved Samhita. Vedic Mathematics helped to transform the universal set of knowledge as a speaking language and in the process it itself as well transformed as such and assimilated its identity into the Vedas.

Within Vedas, all discipline of knowledge transform their identity and get assimilated into the single discipline of organization of knowledge on geometric formats. Vedic geometry and mathematics as such help us to work out these formats.

VIII
Glimpses of Vedic Mathematics

[Authored (1990) by Prof. Krishnaji, Prof. of Physics and Ex-Pro-Vice Chancellor, Allahabad University, Allahabad, India.]

Vedic mathematics was lost to the modern world over several thousand years. Fortunately, the renowned scholar Jagadguru Swami Bharti Krishna Tirthji Maharaj, Shankracharya of Sharda Peeth and Goverdham Math discovered 16 Vedic Sutras and 13 Upsutras in the Parishistha of Atharvved. After continuous Sadhana and hardwork, he was able to decode them and get wide ranging mathematical principles and applications from them. He found that these Sturas cover all aspects of mathematical science in depth. He wrote sixteen volumes on Vedic mathematics covering all aspects of these sutras in depth. Unfortunately, the manuscript was lost which fact was finally confirmed in 1956. During his last days when his health was failing and his eyesight was weak, he rewrote from his memory an introductory account of the subject. He attained Mahasmadhi in 1960. Later in 1965 a book entitled "Vedic Mathematics" was published. This book is edited by Dr. V.S.Agarwala with the help of Smt. Manula Trivedi on the basis of manuscript and notes left by Swamijee. This book is therefore, the first book to give a glimpse of Vedic mathematics and to reestablished the subject. The book attracted the attention of several Indians and foreigners. Dr. Narinder Puri of Roorkee University took up the cause of Vedic Mathematics in right earnest. His Holiness Maharishi Mahesh Yogi spread the message of Vedic mathematics throughout the world by sending Dr. Puri to several countries and making it possible for many schools all over the world start teaching Vedic mathematics. A former minister of Human Resource Development, in the Government of India, Sh. P.V.Narasima Rao made a reference in Parliament about the work on Vedic mathematics which created a wider interest in the nation. Rashtriya Ved Vidya Pratishthan started looking into the matter and appointed a committee on

Vedic Mathematics. During the same period a controversy arose about the source of these 16 Sutras and the level of mathematics contained in them. One view was that these Sutras do not have a Vedic source and that they deal only with simple arithmetical operations.

In the meantime, Dr. Sant Kumar Kapoor, inspired by integrated impact of the Ganita Sutras, started working on basic aspects of the mathematical structure of Vedic literature. He got several clues from Ganita Sutras, Maheshwara Sutra and many other Vedic Mantras. He was inspired by the divine blessing from Yogiraj Sri Sripad Babaji of Vrindavan and His Holiness Maharishi Mahesh Yogiji at Maharishi Nagar. He has been awarded Ph.D. degree by Kurukshetra University, in December 1990 on his thesis titled "Mathematical Basis of Vedic Literature". His work shows that the potential of Ganita Sutras is much higher than what appears in the book "Vedic Mathematics" and the mathematical structure of Ganita Sutras is similar to that of Saṃved indicating that Ganita Sturas belong to Vedic family. His work shows that the Devnagri alphabet and the Vedic scriptures are mathematically organized. Perhaps a more important aspect of his work in the existence of Real Geometric Spaces of higher dimensions in the form of a continuum. A comprehensive description of his research is given in the next section. Dr. Sant Kumar Kapoor is presently working as a Visiting Professor in the Indian Institute of Maharishi Vedic Science & Technology which is a part of Maharishi Ved Vigyan Vishwa Vidyapeeth, Maharishi Nagar. Three research students are working in our institution on the structure and properties of fourth, fifth and sixth dimensional spaces.

Vedic Mathematical Research done by Dr. S. K. Kapoor

Studies reveal that knowledge and organisation of knowledge are two distinct disciplines. The thesis of Dr. S.K. Kapoor is substantively covering the organisational aspect of Vedic knowledge. The main results of the thesis and his subsequent research can be summed as under:

Real 4 and higher dimensional spaces do exist.

Real 4 and higher dimensional spaces were not only known to the Vedic seers but those were precisely availed by them for organisation of vedic knowledge

Vedic alphabet format is a precise setup of 4-space. The alphabet letters of Devnagri script, their forms formulations, combinations and rules etc. are precisely mathematical.

Sakala Rigvedic Samhita is the oldest scripture of mankind. The organisational format of this Samhita is precisely that of real 6-space.

Vedic knowledge accepts a measuring rod constituted by regular bodies of the first six real dimensional spaces which admit formulations as:

Interval	Square	Cube	Hypercube-4	Hypercube-5	Hypercube-6
		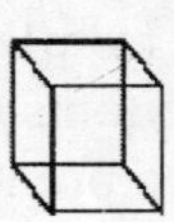	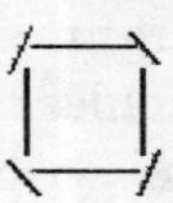		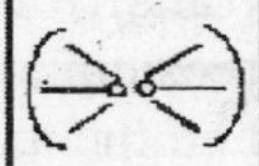
1-space body	2-space body	3-space body	4-space body	5-space body	6-space body

The organizational format of Srimad Bhagwad Gita emerges to be of prime importance as it is parallel to the organizing power of the knowledge content of the scripture. This format tallies with the format of human body on the one hand and the sun on the other hand as of real 6-space.

The Trinity of Gods namely Brahma, Mahesh, Vishnu are the overlords of real 4, 5 and 6 spaces respectively.

The geometrical continuum expressed as manifestation layers of 4 folds of consecutive dimensional spaces contents. These folds of the nth manifestation layer can be represented as under:

First Fold	**Second Fold**	**Third Fold**	**Fourth Fold**
Dimension	Frame	Domain	Origin
(n-2) space content	(n-1) space content	n-space content	(n+1) space content

The transcendence from one manifestation layer to another manifestation layer giving rise to the following (five steps) chain reaction or five steps which are possible within the setup of Panch Mahabhut.

Manifestation Layer	**Dimension**	**Frame**	**Domain**	**Origin**
Nth	Space fold (n-2)	Space fold (n-1)	Space fold n	Space fold (n+1)
(n+1)th	Space fold (n-1)	Space fold n	Space fold (n+1)	Space fold (n+2)

Transition from one space to another space is to be had in terms of unlocking of the seals of the origin points of all the four folds of the manifestation. The modern mathematical models of transition from straight line to plane deserve serious reexamination. In particular the axioms of space filling curves and the axioms of 'one' without a predecessor deserve close scrutiny as their rationale emerges to be without basis.

The role of real numbers additive group (R,+) and real numbers field (R,+,×) with reference to straight line deserve to be differentiated.

The plane deserves to be studied as four geometrically distinct quarters. One faced plane and two faced plane are two distinct geometrical setups and they deserve to be taken up as such.

The concepts of origin and dimension are two concepts with respect to which the modern geometrical models are not up to date. These two concepts deserves to be studied in detail as transcendence to the higher dimensional spaces is possible only in terms of their understanding.

Human body is a compactified phenomenon of multi-layer physiological existence. The start with state of existence is that of waking state which is parallel to the expression of 1-space as dimension into 3-space domain. Sequentially, the existence phenomenon unfolds until seventh state of consciousness which would be corresponding to the 7-space as dimension into 9 space domain. The origin point of the 6-space, being the 7-space setup, the human body, geometrically, turns out to be hypercube-6 and this would explain how the primordial sound, the planetary effects, the Yajna oblations etc. operate and precisely influence the individual existence patterns.

Srimad Bhagwad Gita is one such scripture whose organisational format precisely workout for us the structural set up and frames of the 6-space. The study zone of Srimad Bhagwad Gita can be worked as under:

Srimad Bhagwad Gita
Study Zone

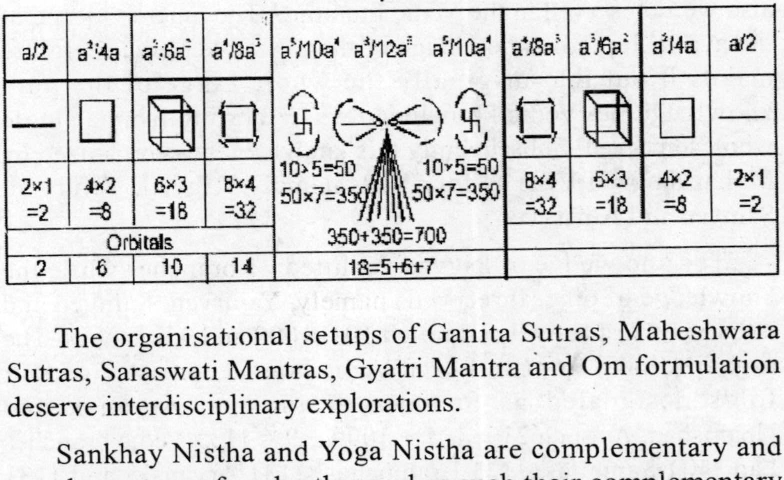

$a/2$	$a^2/4a$	$a^3/6a^2$	$a^4/8a^3$	$a^5/10a^4$	$a^6/12a^5$	$a^5/10a^4$	$a^4/8a^3$	$a^3/6a^2$	$a^2/4a$	$a/2$
—										—
2×1 =2	4×2 =8	6×3 =18	8×4 =32	10×5=50 50×7=350	350+350=700	10×5=50 50×7=350	8×4 =32	6×3 =18	4×2 =8	2×1 =2
Orbitals										
2	6	10	14	18=5+6+7						

The organisational setups of Ganita Sutras, Maheshwara Sutras, Saraswati Mantras, Gyatri Mantra and Om formulation deserve interdisciplinary explorations.

Sankhay Nistha and Yoga Nistha are complementary and supplementary of each other and as such their complementary nature and supplementary nature deserve to be distinguished well. Non-differentiation of the same is bound to deprive us of most of the results in specific forms.

IX
Multi-Dimensions of Time and Space in Manasara

[This Article of Dr. S. K. Kapoor is published in January 2002 issue of Vedic Vastu Sandesh, Indore, India.]

PART I: VEDIC WISDOM

VEDIC WISDOM

Vedic wisdom is lively in the consciousness of living saints. It is also well preserved in the Vedic literature. The inner evidence of the available Vedic knowledge of the available Vedic literature makes it out that originally the whole range of the pure knowledge i.e. Vedic knowledge was vibrating from a single wholesome Ved. Subsequently this knowledge was organized by Maharishi Ved Vyas as four Veds namely, Rigved, Yajurved, Samved and Atharvved.

The knowledge of Rigved admitted 21 branches while the knowledge of other three Veds namely, Yajurved, Samved and Atharvved respectively admitted 101, 1000 and 9 branches. The organizational format of each Vedic branch consisted of four folds designated as Samhita, Brahmana, Aryanak and Upanishad. As such 21 + 101 + 1000 + 9 = 1131 Vedic branches had 1131 Samhitas, 1131 Brahmanas, 1131 Aryanaks and 1131 Upanishads. These 1131x4 = 4524 scriptures together came to be known as Vedic wisdom. In addition corresponding to each Ved, as applied value of the pure knowledge of the Ved is Upved. The four Upveds are Ayurved, Dhanurved, Gandharvved and Sthapatyaved. Our present day mathematics, science and technology come within the range of Sthapatyaupved. Manasara is one such scripture of Sthapatyaupved.

MANASARA: SCRIPTURE OF STHAPATYAUPVED

Sri Prasana Kumar Acharya had done a wonderful job of reconstructing the text of Manasara and preparing its translation in English and by drawing the plates.

The scripture begins with the prayer to Lord Brahma, the creator, the supreme and ends with the chiseling of third eye of the idol of Lord Shiv. The broad organization of the scripture is of the range of 70 chapters with first eight chapters constituting a primary group-I. The next ten chapters i.e. chapter 9 to 18 constitute a primary group-II. Then comes the central part of the Manasara. Chapter 19 to 30 cover single story building to 12 storied buildings. These 12 chapters constitute a central group-I. Next 20 chapters i.e. chapters 31 to chapter 50 cover central group-II and with it the subject of architecture as such is completed. Then follows the subject of sculptures. The science of sculptures covered in chapters 51 to 70 can be organizationally divided in two sculptural groups. Chapters 51 to 65 constitute sculptural group-I and chapters 66 to 70 constitute sculptural group-II.

SPACE TIME FRAME

The above topical division of the text has an organizational message of great importance as in terms of it we may reach at the geometric format of the organization of the knowledge of the scripture.

As the scripture begins with the prayer to Lord Brahma, the overlord of real 4-space and ends with the chiseling of third eye of Lord Shiv, the overlord of real 5-space, therefore, geometric format at the dimensional level is bound to be spatial with the flux of time being solid.

In short, the space time frame at the dimensional level is going to be E^2 (space) × E^3 (time). In terms of this dimensional order we shall be manifesting working geometric domain within the space time frame $(E^2)^4$ (space) x E^3 (solid time). Here (today) we shall be concentrating upon the concepts and comprehensions of dimensions of space and time in Manasara. In a way, we shall be taking up the topic of space, time and space time in the light of Vedic wisdom. In the context, it may be relevant to note that the modern thought, mathematics, science and technology is speculating the general space time frame as E^3 (space) × E^1 (linear time).

The modern thought, mathematics, science and technology centre around linear dimensional reality but the Vedic systems avail multidimensional reality. The studies of the organizational formats of various Vedic scriptures reveal that higher dimensional geometric formats are being availed to organize the pure knowledge. Illustratively, we may take the case of the oldest book of mankind namely, Sakla Rigved Samhita.

RIG VED SAMHITA

Fortunately Rigved Samhita is intact with us from first syllable to the last syllable and as the tradition goes, the whole range of Vedic knowledge is lively in this scripture of 432000 syllables, out of which 397265 syllables are manifest text while remaining 34735 syllables go deep as organizational format of the text and as such remain un-manifest.

For the present, we may accept it an axiom that knowledge and organization of knowledge are two distinct aspects of knowledge. Being scriptural text, we get the organized knowledge and as such both organization format and the text are to be accepted as the knowledge content of the scripture. It is like a truck with goods yielding weight of the truck as well as of the goods loaded in the truck.

Organisation of Rigved Samhita

Total knowledge contents	432000 Syllables
Manifest text	397265 Syllables
Mandals	10
Ashtaks	08
Chapters	64
Anuvaks	85
Suktas	1028
Vargas	2024
Richas	10552

Mathematical Basis

Mathematical basis of the organisational format of the Rigved Samhita reveals that the Vedic knowledge is organised on geometric format of real 6-space. It admits 4-space in the role of dimension while modern thought, mathematics, science and technology centre around 3-space reality and as such Veds are invincible fort for the modern mind. As such, we have to learn and understand the Vedic wisdom. For this, we have to re-examine the rationale and basis of our axioms and postulates for accepting the reality as linear dimensional one. It is only by approaching the Vedic knowledge, the Vedic way, that we may have real bliss of Vedic wisdom.

PART II: HIGHER DIMENSIONAL REALITY MATHEMATICAL BASIS OF VEDIC LITERATURE

Studies of mathematical basis of available Vedic literature reveal that 4 and higher dimensional reality was not only known to the Vedic seers rather the great use thereof was made by them for organization of pure knowledge.

Let us concentrate upon the nature of space around us. We can see that straight line is a track of a moving point while plane is a track of a moving (straight) line. Likewise, moving plane shall be creating solid space. More precisely, if we reinvestigate this phenomenon with the help of a point, interval, square and cube, we shall be noticing that moving point accepts line as its track, moving interval accepts square as its track and moving square accepts cube as its track. This would pose a question: What is the nature of the track of moving cube/solid /3-space body/ 3-space setup or in the general, 3-space itself?

HYPERCUBES 4, 5, and 6

To answer the same let us mathematise this situation as:

Taking point as a representative regular body of 0 space, interval, square and cube respectively as representative regular bodies of 1, 2 and 3 space, we may conclude that moving 0-space (body) accepts 1-space format, moving 1-space (body) accepts 2-space format; moving 2-space body accepts 3-space format, and as a logical consequence the moving 3-space body shall be requiring 4-space format. And in general, moving n-space body shall be requiring (n+1) space format. For convenient handling.

We may define and designate four and higher dimensional bodies in continuation of interval, square and cube as hypercubes. To be precise hypercube 4 shall be a representative regular body of 5-space and so on. Further to facilitate comprehension of main properties of hypercube and for symbolic representation of the set of properties synthesizing hypercube it would be desirable to have suitable symbols for them, particularly for hypercube 4, 5 and 6 for reaching at the concepts and comprehension of multi-dimensional spacetime frame being availed for organization of the knowledge of Manasara.

Let these 3 special symbols of hypercube 4, 5 and 6 be as:

Interval	Square	Cube	Hypercube-4	Hypercube-5	Hypercube-6
1-space body	2-space body	3-space body	4-space body	5-space body	6-space body

To have these symbols in continuity of the geometric setup of interval, square and cube, we have to first comprehend these geometric setups. As such, let us have a close look at the geometric setup of interval, square and cube. If we have close look at the geometric setup of interval, square and cube, we may comprehend that interval has length (A^1) and 2 boundary points ($2A^0$) Square has area (A^2) and 4 boundary line ($4A^1$) and cube has volume A^3 and 6 boundary surfaces ($6A^2$).

A^1:$2A^0$, A^2:$4A^1$ and $A^3.6A^2$ suggest a common formulation $A^n.2nA^{n-1}$, n=1, 2, 3.

This formulation would hold for all values of n. In particular, for n = 4 5 and 6 we shall be getting A^4:$8A^3$, A^5:$10A^4$ and A^6:$12A^5$. The geometric message is that boundary of 4-space body is constituted by 8 cubes while the boundary of 5-space body is constituted by 10 hypercubes-4 and so on.

The boundary components of interval, square, cube and hypercubes 4, 5 and 6 are in the ratio 2:4:6:8:12.This would help us comprehend, appreciate and have the symbols of hypercube 4, 5 and 6 in continuity of and in that sequential order, hypercube 4, 5 and 6 with boundaries as:

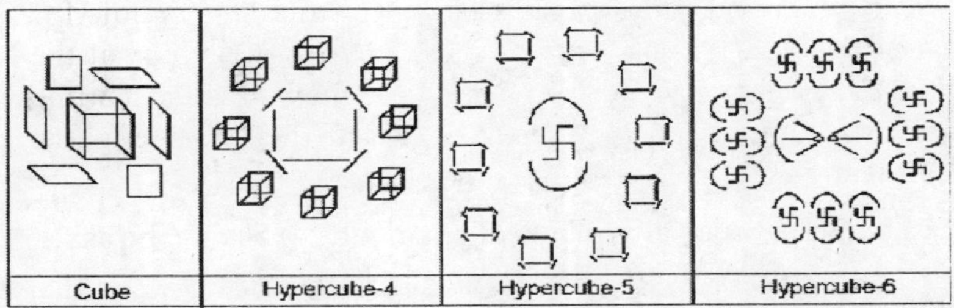

SHAD CHAKRA FORMAT OF HUMAN BODY

Further it would help us comprehend and appreciate the Upanishad knowledge regarding the Shad Chakra format of human body as that:

अथ बहिर्लक्षणाम्। नासिकाकग्रे चतुर्भि षड्भिरष्टभि. दशभि. द्वादशभि. कमात्।

The external characteristics, ahead of tip of nose, are four, six, eight, ten and twelve in that sequence.

These, as is evident are parallel to the sequence an order of boundary components of square, cube, hypercubes-4, 5 and 6 as being four (lines), six (surface plates), eight (solid components), ten (hyper solid-4 component) and twelve (hyper solid five component) respectively.

With this the Shad-Chakra format of human body accepting Sathapatya measuring rod constituted by representative regular bodies of 1 to 6 Space makes the human existence as within human frame of Shad Chakra format is of the order and as of format which is to run parallel to the manifestation and de-manifestation processes of flow of Jyoti from orb of the Sun through its rays.

CONCLUSION

The conclusion of all conclusions comes to be that Vedic systems have successfully unified whole range of knowledge as a single discipline accepting Sathapatya measuring rod.

Section II

GANITA SUTRAS' GEOMETRIC FORMATS

LORD GANESHA,

गणेश=	गण + ईश
GANESH	=GANA+ ISH
गणित =	गण + इत्
GANITA	=GANA + IT

The lord of ganas, with whose grace sadhkas perfect intelligence to learn ganita by learning to remove the stitching thread of ganas.

Ganita Sutras Text

01. एकाधिकेन पूर्वेणः	*Ekadhiken Purvena*
02. निखिलं नवतश्चरमं दशतः	*Nikhilam Navatascramam Dasatah*
03. ऊर्ध्वतिर्यग्भ्याम	*Urdhvatiryagbhyam*
04. परावर्त्य योजयेत्	*Paravartya Yojayet*
05. शून्य साम्यसमुच्चये	*Sunyam Samyasamuccaye*
06. (आनुरूप्ये) शून्यमन्यत्	*(Anurupye) Sunyamanyat*
07. संकलनव्यवकलनाभ्याम्	*Sankalana-vyavakalanbhyam*
08. पूरणपूरणाभ्याम्	*Puranapuranabhyan*
09. चलनकलनाभ्याम्	*Calana-kalanabhyam*
10. यावदूनम्	*Yavadunam*
11. व्यष्टिसमष्टि	*Vyastisamastih*
12. शेषाण्यङ्.केन चरमेण	*Sesnyankena Caramena*
13. सोपान्त्यद्वयमन्त्यम्	*Sopantyadvyamantyam*
14. एकन्यूनेन पूर्वेण	*Ekanyunena Purvena*
15. गुणितसमुच्चयः	*Gunitasamuccayah*
16. गुणकसमुच्चयः	*Gunaksamuccayah*

*****

Ganita Sutra-1

एकाधिकेन पूर्वेण:

Ekadhiken Purvena

One more than before

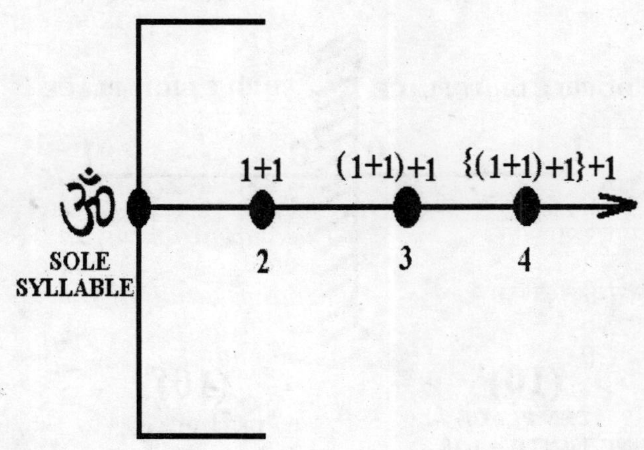

OM (ॐ) IS THE SOURCE FORMULATION. AS SUCH GANITA SUTRA-1 PROCEEDS FURTHER WITH THE ORDERING RULE OF ONE MORE THAN BEFORE PRESUMING THE EXISTENCE AND AVAILABILITY OF THE SOURCE FORMULATION OM .

Ganita Sutra-2

निखिलं नवतश्चरमं दशतः

Nikhilam Navatascramam Dasatah

All from nine and last from ten

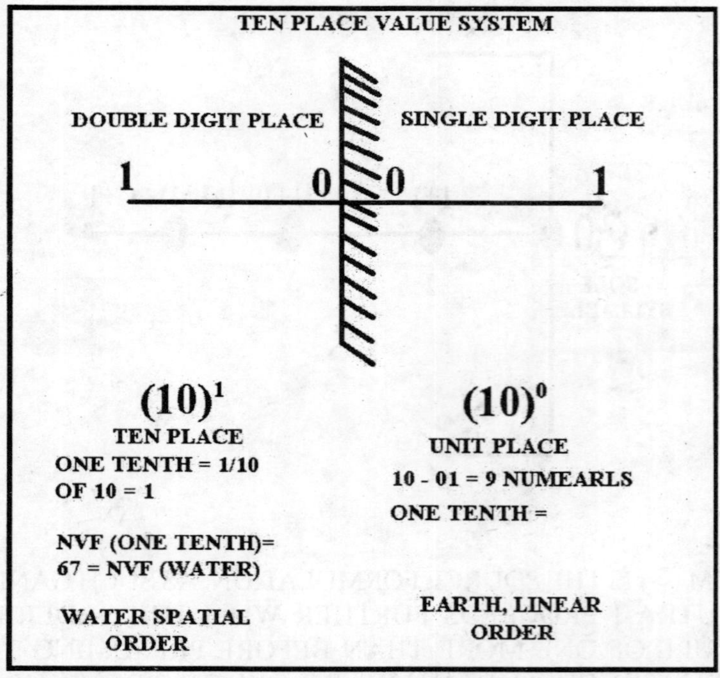

PROGRESS: Sutra-1 provides format for counting; Sutra-2, a step ahead provides format for place values.

Ganita Sutra-3

ऊर्ध्वतिर्यग्भ्याम्

Urdhvatiryagbhyam

Vertically and crosswise

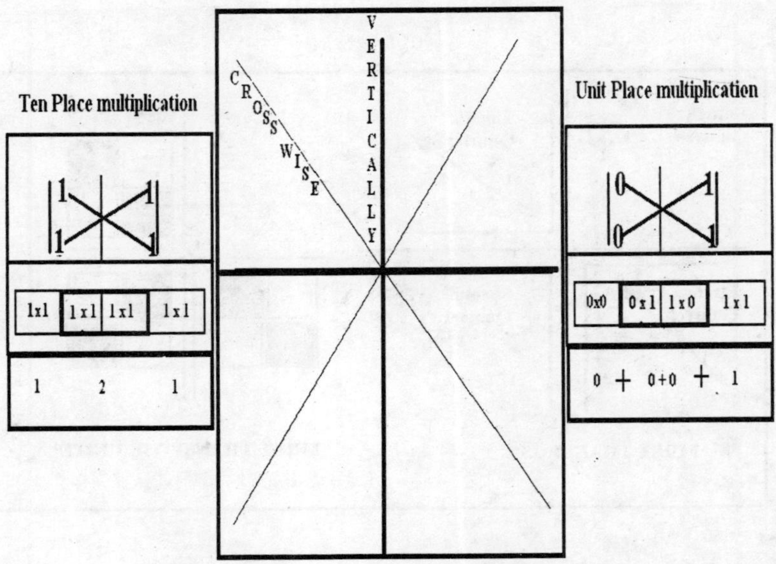

PROGRESS: Sutra-1 format adds counting order. Sutra-2 format adds place value order with the help of mirror.Sutra-3 format adds structural processing (along artifices of numbers) with the help of axes.

Ganita Sutra-4

परावर्त्य योजयेत्

Paravartya Yojayet

Transpose and unite

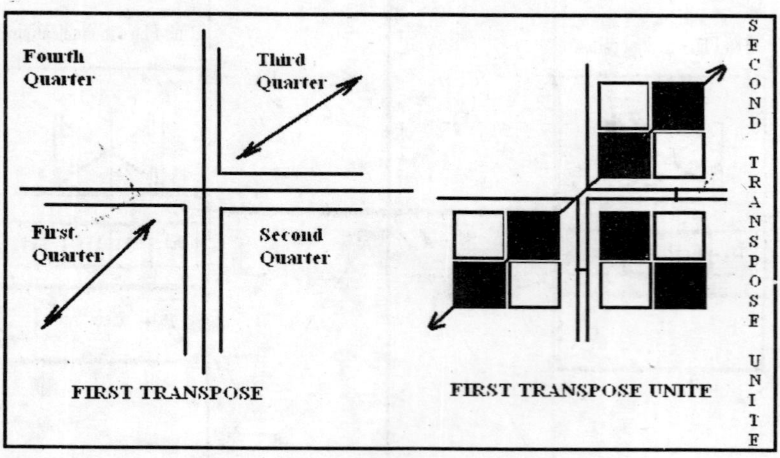

PROGRESS: Sutra-1 avails artifice of whole number –1, as single digit expression parallel to the format of 1-Space. Sutra-2 avails artifice of whole number-2 as double expression parallel to 2-Space format. Sutra-3 avails artifice of whole number-3 as triple digit expression parallel to 3-Space format. Sutra-4 goes a step ahead and avails pair of double digit expressions parallel to spatial order of 4-Space and parallel to artifice of whole number –4 of unique features 2 + 2 = 2 x 2 -2 x -2 because of which with second transpose diagonals unite.

Ganita Sutra-5

शून्य साम्यसमुच्चये

Sunyam Samyasamuccaye

If the Samuccya is the same it is zero

I

Sutra-4 Format avails artifice of whole number – 4 parallel to spatial order of creator's Space (4-Space). Because of it quarters get potentialized as of NVF (QUARTER) = 17 + 21 + 1 + 18 + 20 + 5 + 18 = 100, which further being of quarters of AREA, NVF (AREA) = 1 + 18 + 5 + 1 = 25.

These quarter areas, as 100 and 25 = 125 = 5 x 5 x 5, emerge as of solid order. It is sequential order for affine format of 5 x 5 arrays of artifices "1".

It is spatial order of affine format transforming as dimensional order with rows and columns of matrix 5 x 5 instead of remaining constant values as 1, go on increasing as 1, 2, 3, ...

It is this solid order as dimensional order takes to transcendental world (5-Space). The boundary of hyper cube-5 is constituted by 10 x 8 = 80 solids/cubes which is of the order of NVF (CREATOR) = 80.

Further NVF (SOLID) = NVF (LINEAR) = NVF (DOUBLE) makes NVF (VOID) = 50 = 10 x 5 = 50 coordinates fixation of boundary of hyper cube-5/5-Space.

With 5-Space as origin of 4-Space, and 4-Space being of spatial order, makes hyper cube-5 structurally as of pair of setups each being reflection image of other.

It is because of it, both setups of hyper cube-5 emerge to self-sustained setups. It is because of it that the sequential logic which takes to hyper cube-5 in five steps (as of sutras 1 to 5) in fact covers only up till one part of hyper cube-5. And, as sixth step, the processing is to begin afresh.

It may be refreshing to refer as that chapter-1 and chapter-6 of Shrimad Bhagwad Geeta are both of 47 shalokas ranges. Still further it also would be refreshing as that chapter –5 of Shrimad Bhagwad Geeta and Chapter-13 Shrimad Durga Sapatsati are both of 29 Shalokas ranges.

The artifice 47 is NVF of monad. The artifice 29 is NVF of black. This that way covers first phase of organization of Ganita Sutras.

II

Affine Matrix 5 x 5 And Sequential Matrix 5 x 5

1 1 1 1 1	Sum 25	1 2 3 4 5	Sum 125
11 1 1 1	= 5 x 5	2 3 4 5 6	= 5 x 5 x 5
1 1 1 1 1	= Square	3 4 5 6 7	= Cube
1 1 1 1 1	=	4 5 6 7 8	
1 1 1 1 1		5 6 7 8 9	

III

Creator

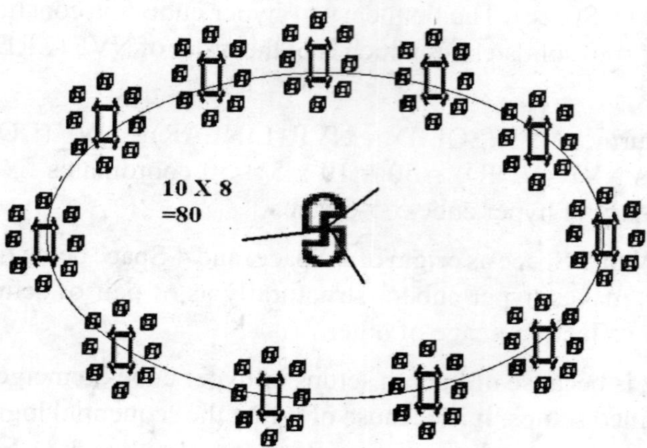

IV

Northern Hemisphere and Southern Hemisphere split up is there because of the pair of parts of 5-Space emerging within 4-Space at its origin.

Ganita Sutra-6

(आनुरूप्ये) शून्यमन्यत्

(Anurupye) Sunyamanyat

If one is in Ratio the others is Zero

I

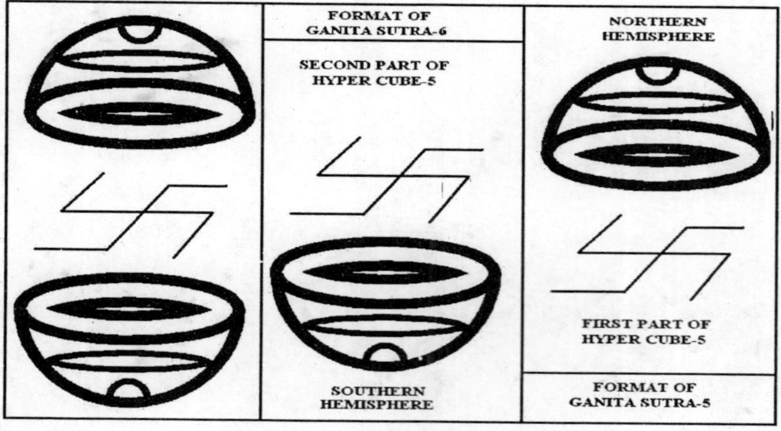

The transition from format of Sutra-5 to Sutra-6 can be had parallel to the transition from the artifice of 5 to artifice of 6 as 5 = 2 + 3 and 6 = 2 x 3.

It is parallel to transition in terms of Ganita Samuchya to Gunak Samuchya. The middle 4-Space also preserves change of orientation parallel to 2 + 3 and 3 + 2.

The composition component Anurupye/ following form, parallel symmetry like that of reflection pair indicates the special feature of this organization.

II

The progress of Sutra-1 is to manifest counting format availing 1-Space, as positive orientation, as artifice of 1 to express single digit values.

Sutra-2 manifests place value format availing 2-Space as artifice of 2 to express double-digit values availing mirror for pairing of artifices.

Sutra-3 manifests format for arithmetic operations for structural organizations on format of 3-Space availing axes to simultaneously handle linear, solid and double orders.

A step ahead Sutra-4 provides break through from linear order to spatial order for the chase of the plane, quarter by quarter.

Sutras-5 & 6 by chasing quarters as well quarter by quarter, attain transition from spatial order to solid order by sequencing the affine planes and thereby transforming area square values into volume cube values.

With this quarter by quarter chase of quarters, there emerges organization format of sixteen sub quarters each of value 125 and thereby the total value as 2000 manifests a pair of 1000 values setups which together as upper part and lower part emerge as transcendental formats for Sutras 5 and 6 respectively.

Ganita Sutra-7

संकलनव्यवकलनाभ्याम्

Sankalana-vyavakalanbhyam

By addition and by subtraction

NVF (ADDITION) = 76 = NVF (MINUS), and the artifice 76 with six at unit place and seven at ten place, as a format firstly is a step ahead of the artifice of six which takes up till the format of Sutra-6.

Further parallel to the spatial order of 4-Space and pair of parts of 5-Space at the seat of origin within 4-Space and additional subtraction as of opposite orientation simultaneously avail the manifestation format of 1 – manifestation layer with + 1 space as domain fold and –1 Space as dimension fold.

With this, 0-Space becomes lively in between devoid of any such orientation features.

NVF(DEVOID)=59
= NVF(LINEAR)
= NVF(SOLID)
= NVF(DOUBLE),

really makes the whole setup capable of being handled simultaneously as affine setup despite Orientations, sequences and dimensions, mirrors, axes as well as the fire-edges with fire as the third element and responsible for solid order for the transcendental world (5-Space) as NVF (FIRE EDGE) = 59.

It is this feature because of which the whole range of knowledge permits organization and chase on geometric formats as a single discipline of knowledge and the gentle arithmetic operations like addition and subtraction emerge to be the most basic operations in terms of which only it becomes possible to perfect the intelligence as of the order of nature.

NVF (SUBTRACTION)
= NVF (SUB) + NVF (TRAC) + NVF (T-ION)
= NVF (NEW) + NVF (NEW) + NVF (NEW + 16).
NVF (SIXTEEN) = 6 x 16
= 6 x NVF (SIXTEEN)
= 6 x 6 (NVF (SIXTEEN)
= 6 x 6 x 6 x NVF (SIXTEEN) and the sequence continues.

This 6-Space (SIX-TEEN) as the origin sequentially manifests pentagon within pentagon and coordination 6 to 5 as 6 – 1 and from 5 to 6 as 5 + 1 makes the functional format of Sutra.

II

FUNCTIONAL FORMAT

NVF (FUNCTIONAL) = 6 + 21 + 14 + 3 + 20 + 9 + 15 + 14 + 1 + 12 = NVF (NEW FORMAT) = 115 = 5 x 23 = 5 x NVF (END). It leads to Pentagon having five ends/corners.

The following figure be chased to see the richness of Pentagon creating another pentagon within it and sequence to envelop the center/origin is available. Sequential mapping of Pentagons of its origin and the flow from center sequentially manifesting the pentagons with common center is the functional format, format to new format and from new format back to the old format with NVF (FORMAT) = NVF (OLD) + NVF (NEW) is the beauty of this affine state functional format of nature with Sun as the center and the transcendental world as the domain of recycling and self-referral processing setup.

The solid order and parallel to it the artifice of three made available by the transcendental world (5-Space with Pentagon as functional format) and hyper solid order and parallel to it the artifice of four made available by Sun as center is there as "One" with NVF (ONE) = 34 and 43 as its reflection pair is NVF (BOOK).

As such One Book and Book One, are there to give us the insight about this functional format of Sutra –7.

Ganita Sutra-8

पूरणपूरणाभ्याम्

Puranapuranabhyan

By the completion or non-completion

The pair of orientation accepted by close interval, makes it a manifestation layer of four folds (-1 Space as dimension, 0-Space as boundary, + 1 Space as domain and 2-Space as origin fold).

The close interval being a manifestation layer with 1-Space as domain fold and further as:

NVF (LINEAR) = NVF (SOLID) and also as the same is as well equal to NVF (DOUBLE) and further it also being equal to NVF (FIRE EDGE), so when it is attempted to be cut into pair of parts, the monad/seed grain of domain which shall be coming within the blades of scissors, the same shall be the third independent part.

As such when the close interval is to be taken as of two parts only, then the third part would be included in first or the second part and with it both the parts shall be becoming unequal, as much as that one of it shall be closed interval and other half closed interval.

It may be taken as that one of its parts would be of full structures of close interval while the remaining would be incomplete as much as that it would be divide of one monad. NVF (ONE MONAD) = 34 + 47 = 81 = NVF (SQUARE), the origin fold of this manifestation layer.

This as such is the truth of the well known Fermats Last Theorem. The conceptual term Purnam accepts translation as "FULL/complete" like close interval.

Accordingly Apurnam is "incomplete", like second part that is half open interval.

Close Interval as of two parts

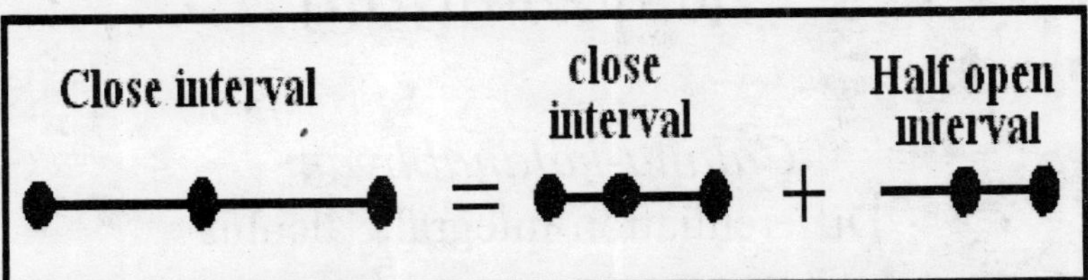

Ganita Sutra-9

चलनकलनाभ्याम्

Calana-kalanabhyam

Differentiation-integral Calculus

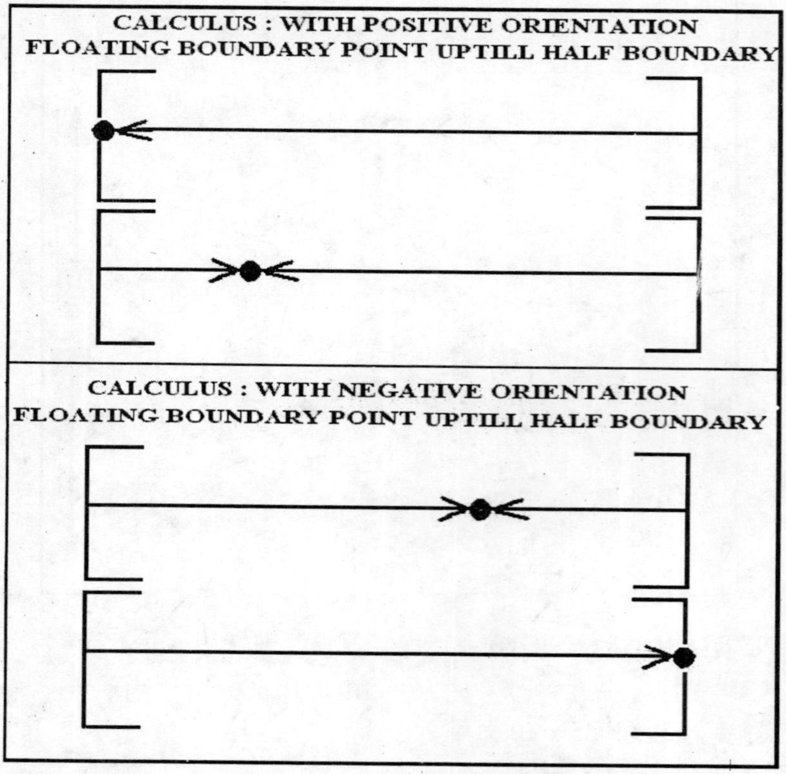

CALCULUS : WITH POSITIVE ORIENTATION
FLOATING BOUNDARY POINT UPTILL HALF BOUNDARY

CALCULUS : WITH NEGATIVE ORIENTATION
FLOATING BOUNDARY POINT UPTILL HALF BOUNDARY

Ganita Sutra-10

यावदूनम् Yavadunam

By deficiency (of double of it)

Sanskrit Composition

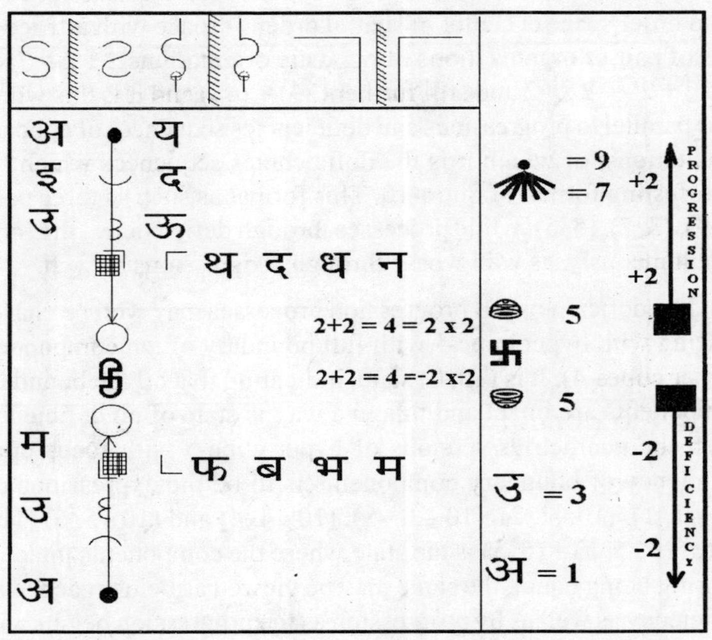

NVF (PROGRESSION)

= 16 + 18 + 15 + 7 + 18 + 5 + 19 + 19 + 9 + 15 + 14

= 155 = 3 x 5 x 10

= 38 + 38 + 38 + 41

= NVF (FIRE) + NVF (FIRE) + NVF (FIRE) + NVF (lock)

As such "fire" trimonad is to take to affine/lock. The Sun (54) beam (21) end (23), with 10 units, makes reflection pair (38, 83), as 38 + 1 and 83 + 10. NVF (TEN) = 39 = NVF (ANGLE) and 93 = 31 + 31 + 31 = NVF (CUBE) + NVF (CUBE) + NVF (CUBE), this way takes to other side of Sun through its beam end **angle**.

NVF (DEFICIENCY) = 4 + 5 + 6 + 9 + 3 + 9 + 5 + 14 + 3 + 25 = 83 Follow (83, 38) is a generic NVF reflection pair. The artifice 83 is as of NVF of GENERIC to FOLLOW. This is a chase of 3-Space cut as eight octants. In other orientation it is eight octants synthesizing 3-Space. The artifice 38 is as of NVF of generic "FIRE". The cut of 3-Space as eight octants is to release 4-Space at the origin. The further transcendence is to lead to 5-Space of solid order as enveloped within ten hyper cubes 4. It is 5-Space as solid order space at center of spatial order 4-Space with artifice of 4 as of pair of organizations of opposite orientations 2 + 2 = 2 x 2 and 2 + 2 = –2 x –2 and still further (24) = (42) and it is this which runs parallel to progressions and deficiencies sequences of opposite orientations, of which it is the deficiencies sequences which are manifesting format of Sutra-10. This format as such as three pairs (1, 9), (3, 7), (5, 5), while processes through deficiencies, the same simultaneously, as well works through progressions as well.

The deficiency and progression processes may well be chased starting with hyper cube-5 with full boundary of ten components (hyper cubes-4). It is (10, 0) state, indicating that all the boundary components are intact and this in a way is state of no deficiency. Then the geometries/versions of hyper cube-5 with focus upon deficiency of boundary components is to be the expressions as: (10 –1, 1), (10 –2, 2), (10 –3, –3), (10 –4, 4) and (10 –5, 5). This state (10 –5, 5) = (5, 5) is the state where the components intact or missing being equal, the same may be viewed as being reached by deficiency as well as by progression as the progression begins with the starting position as when whole of the boundary having been stripped of and the state being of (0) boundary component which may accept expression as (0, 10) = (10 –10, 10). The progression may work as (10 –9, 1), (10 –8, 2), 10 –7, 3), (10 –6, 4) and (10 – 5, 5) = (5, 5). This double status for (5, 5) state is the grace a state as NVF (GRACE) = NVF (ONE). This, this way makes 11 versions/ geometries of 5-Space. NVF (ELEVEN) = NVF (LIMIT) = 63 with 3 at unit place and 3 + 3 at next, ten value place, in the context may help us have insight about this organization, format and the working process of Sutra-10 through deficiency, in focus and the same by Anurupya/forms/symmetry is to cover the other part as progression.

Ganita Sutras 11-16

11 व्यष्टि–समष्टि vyasti-samastih

Specific and General

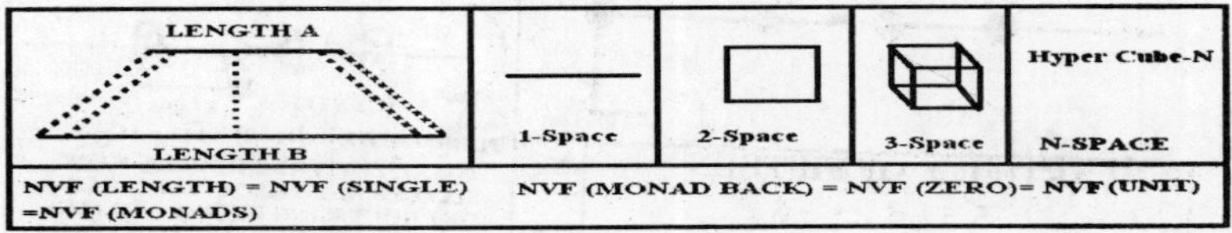

12. शेषाण्यङ्-केन चरमेण Sesnyankena Caramena

The Remainder by the last digit

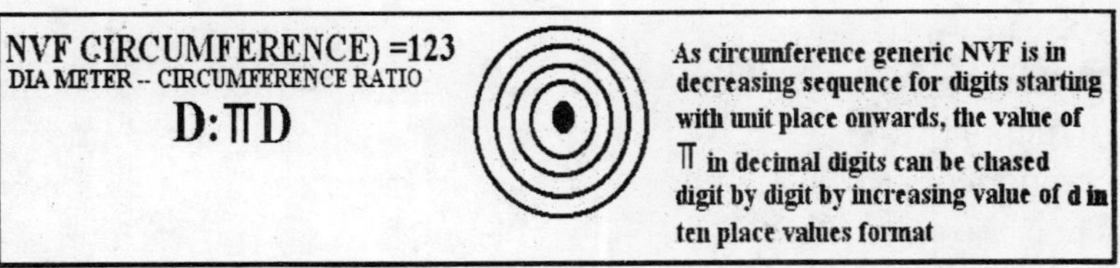

13 सोपान्त्यद्वयमन्त्यम् Sopantyadvyamantyam

The ultimate and twice the penultimate

NVF (ULTIMATE) = NVF (INTERVAL)

NVF (PENULTIMATE) = 136 = 66 + 66 = NVF (SINGLE) + NVF(SINGLE)

DOMAIN BOUNDARY RATIO FORMULATION AN : 2NBN-1.

14 एकन्यूनेन पूर्वेण ***Ekanyunena Purvena***

By One less than the One Before

15 गुणितसमुच्चयः *Gunitasamuccayah* The product of the Sum

16 गुणकसमुच्चयः *Gunaksamuccayah* All the Multipliers

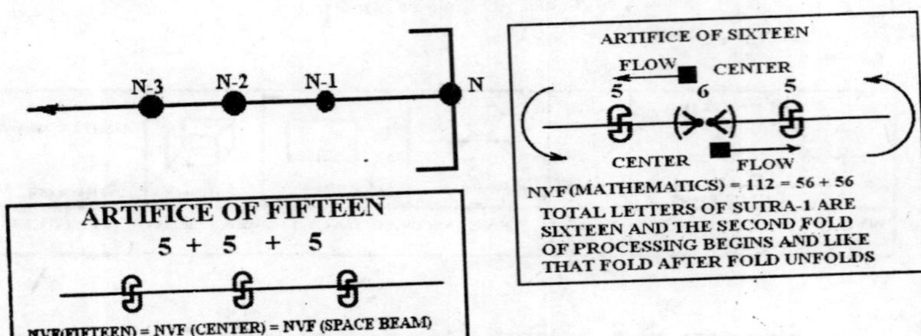

Section III

SPACE BOOK

I
English-Pairing

(A = 1, B= 2, C= 3......... X = 24, Y = 25, Z= 26)

NVF (ENGLISH) = 5+14+7+12+9+19+8= 74

NVF (PAIRING) = 16+1+9+18+9+14+7= 74

NVF (CONE) + NVF (CONE) = 37 + 37 = 74

II
A, An, The

NVF (A) = 1   NVF (AN) = 15   NVF(THE) = 33

Artifices 1, 15 and 33 have their unique properties.

The artifice 1 is unique as that 1 = 1 x 1 = 1 x 1 x 1 and so on.

The feature of artifice of 1 is also like that of monad without parts.

The artifice of 15 is unique as it accepts organization as product of first three odd numerals viz. 15 = 1 x 3 x 5. It as such is the expression of the dimensional order range of the transcendental world (5-Space, which is a solid dimensional order space and solid dimensional order itself is again of linear dimensional order).

NVF (FACE) = 15 further makes this artifice being unique as being the FACE.

Artifice of 33 is unique as it accepts 3 at unit place as well as at next place. This pairing of artifice of 3 has parallel of division of three dimensional frame as a pair of three dimensional frames of half dimensions.

NVF (SEED) = 33 further makes this artifice of 33 being unique as being the SEED.

These distinct NVFs of A, AN and THE may help us have an insight for their distinct application in the English grammar, illustratively as, for a sole part as the part of an article.

III

That, This

NVF (THAT) = 49 = NVF (LORD) and

NVF (THIS) = 56 = NVF (LIGHT) = NVF (FLOW) = NVF (DOMAIN) may help us have insight into the generic classification of words formulations as "THAT & THIS"

IV

One

NVF (ONE) = 15 + 14 + 5 = 34

34 = 1 + 33 = NVF (A SEED).

The generic classification, starting with A SEED, is accepted as ONE.

The artifice 34 with 4 at unit place and 3 at next place makes it a flow as in that direction. The artifice 3 as formulation THREE with NVF(THREE) = 56 = NVF (FLOW) = NVF (LIGHT), as such takes to that order of flow from THAT (LORD, CREATOR/ Presiding deity of 4-Space) to THIS (THREE, FLOW, DOMAIN).

NVF (FOUR) = 6 + 15 + 21 + 18 = 60 = 5 x 12 as sixty coordinates of twelve hyper cubes-5 as boundary components of hyper cube-6 may help us appreciate this flow from the dimensional order of 6-Space (with 4-Space in the role of dimension) to dimensional order of 5-Space (with 3-Space in the role of dimension) and 5-Space itself in the role of boundary of 6-Space.

This, this way, may help us have insight into the organization formats of "ONE" as formulation for coordination of the artifices of the numbers availed parallel to the letters constituting the composition of the word "ONE".

V

Two

NVF (TWO) = 20 + 23 +15 = 58 = 28 + 30 = NVF (THIS) + NVF (BIBLE).

NVF (TWO) = NVF (THIS BIBLE), and 58 = 1 + 56 + 1 = 56 as middle part/middle space, as flow from THAT to THIS and further, 56 = 49 + 7, making available 7 for beyond THAT is to make it out as that the seven streams flow within rays of the SUN is to take even beyond that with NVF (SUN) = 54 = 56 – 2 as middle space is to make Sun as THIS for THAT LORD, which is beyond orb of the Sun and is the *Sapat Rishi Lok* (7-Space), where the artifice of 7 pairs as 77 because of the domain boundary formulation for 7-Space being of the order $A^7 : 14B^6$. It is the pairing of (7, 7) which accepts addition of values along tri-monad within 6-Space because of 1 x 2 x 3= 1 + 2 + 3, and the same along Tetra Monad within 7-Space accepts pairing along the pair of place values as 77. It is the unity state with transcendental world (5-Space) in the role of dimension of 7-Space. Here boundary of transcendental world is of spatial order (2-Space in the role of dimension of 4-Space) and the domain boundary ratio $A^5{:}10B^4$ amounts to organization in terms of ten place value system. With this the pairing being of value 77 is of the order of NVF(CHRIST) = 77.

It is this mathematics of pairing operation which when appropriately applied for transition for triples as a pair of pairings operation, the transition can be attained from SPACE BOOK to SPACE BIBLE along the pairing of artifice of 7. This is the transition from 7 + 7 to 77. The NVF (SPACE) = 44 and NVF (BIBLE) = 30 with 44 –30 = 14 = 7 + 7 with single digit expressions for numerals. For transition from 7 + 7 to 77 organization of pairing of artifices, the first step of transition is to be taken as of transition from single digit formats for numerals to double digit formats for numerals.

The formulation FORMAT with its internal organization as 'for mat' as 'FOR + MAT' takes to their number value formats being NVF (FORMAT) = 73 = NVF (FOR)+ NVF(MAT) = 39 + 34 = NVF (TEN) + NVF (ONE) = (10, 01).

The combined expression for (10, 01) along line may be as:

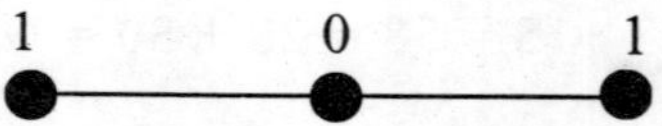

This pairing, (10, 01), as pairing operation, formulating as FORMAT and further making the artifices of TEN and ONE as of five units difference while simultaneously (10, 01) being reflection pair and still further 10= 05 + 05 and also 10 – 01 = 09.

The unfolding of boundary of 5-Space as parallel to ten place value system, *Ganita Sutras-1, 2* and *Ganita Upsutra –1 and 2* together as of working rules "one more than before", proportionately, "all from nine and last from ten" and "remainder is that, that remains (after division)", and further NVF (INTERVAL) = NVF (DIVISION) = NVF (ULTIMATE) = 101, all taken together, make out the pairing operation, as of mirror reflection pairing of objects and images with NVF (OBJECT) = 55 = 32 + 23 and NVF (IMAGE) = 35 = NVF (EYE).

This, this way, the objects images pairing with parallel reflection pairs organizations of artifices of numbers, is to help attain transition from SPACE BOOK to SPACE BIBLE.

This transition is to be from God state of consciousness to unity state of consciousness which through the reflection pairing process is to attain transition from 26 letters English Alphabet to 52 letters *Vedic* Alphabet and it is with this attainment of transition the grand unification of Ancient Wisdom lively as organization formats of whole range of scriptures is to blissfully emerge being single discipline of whole range of knowledge from MATTER to FREQUENCIES as NVF (MATTER) = 77 = NVF (CHRIST) and NVF (FREQUENCIES) = 6 + 18 + 5 + 17 + 21 + 5 + 14 +3 + 9 + 5 + 19 = 122 = 61 + 61 = NVF (CHURCH) + NVF (CHURCH) = NVF (CHRIST) + NVF (RANGE).

VI

(Mirror, Content)

NVF (MIRROR) = 91 = NVF (CONTENT)

Artifice of 91 accepts organization as 1 x 1 at unit place and 3 x 3 at next place. This pair of flow lines from dimension to domain is the unique feature of this artifice and also of this generic of number value format formulations like MIRROR and CONTENT.

NVF (ULTIMATE) = NVF (INTERVAL) = 101 and corresponding organization expression of artifice of 101 along interval as of dimonad format with its pair of parts constituting reflection pair with mirror at its middle may help us have insight about the CONTENT as of common generic with MIRROR.

NVF (GENERIC) = 7 + 5 + 14 + 5 + 18 + 9 + 3 = 62 = 31 + 31 = NVF (CUBE) + NVF (CUBE) may further help us have insight about the pair of flow lines being accepted by the artifice 91 in terms of which generic classification is there for the formulations like CONTENT, MIRROR.

VII

(Linear, Solid)

NVF (LINEAR) = 59 = NVF (SOLID) may help us have insight about the reason why the binary operation accepted by modern mathematics is coming in its way while it accepts so constructed real line of modern mathematics being linear and not solid while in fact that has given an obvious slip and as a result modern mathematician have been deprived of the real features of pairing operation which in fact make the said construction as a solid line as here the generic classification makes the equality for the number value formats of formulations LINEAR and SOLID. It is this distinguishing feature of pairing operation of Ancient Wisdom from that of modern mathematician chase in terms of binary operation, which need be focused and comprehended well.

This may be illustrated and chased as generation status of the elements/members, while as affine space as of zero generation and while those are picked one by one, the same as is the rule of *Ganita Sutra-1* become as of 1 – generation status as members of the set. The binary operation makes them of 2-generation status as operated elements. As A = A x 1 automatically acquires 2-generation status so its operated elements status A x A as (A x 1) x (A x 1) makes it of 3-generation status. It is like this it becomes solid/ domain, ready for flow, as light, as frequencies as of NVF 122 = 61 + 61, a pairing for flow from 1 to 6 = 3 + 3, as three pairs of half dimensions.

VIII

The End, Be End, God

NVF (THE END) = 56 = NVF (DOMAIN) = NVF (LIGHT) = NVF (FLOW)

NVF (BE END) = 30 = NVF (BIBLE)

NVF (GOD) = 26 = NVF (Z)

IX

Seed, Space Seed, Seed Space Seed

NVF (SEED) = 33, as division of three dimensional frame as a pair of three dimensional frames of half dimensions.

NVF (SPACE SEED) = 77 = NVF (CHRIST)

NVF (SEED SPACE SEED) = NVF (SEED CHRIST) = 110 = 11 x 10, parallel to eleven versions of hyper cube-5 as eleven geometries of 5-Space.

Also SEED SPACE SEED with SPACE as middle space, is pairing flow from within the middle, from within the center/ origin, origin of domain/ Space/linear 3-Space with spatial 4-Space as origin,

ONE CHRIST, A SEED CHRIST of 33 revolutions of Earth around the Sun and 34 revolutions as the first revolution of Post Christ Era.

As such SPACE BOOK for its transition to be SPACE BIBLE, it is to be a transition from BOOK of NVF 43 to that of BIBLE of NVF 30 with creator's Space to be de-voided of its hyper cube 13. With it there would emerge a solid dimensional transcendental world from origin of creator's Space. This is transition from spatial order of creator's Space to that of solid order transcendental world. With it there to be transition from pairing as of NVF(CONE) = NVF (CONE) to that of NVF (CONE) + NVF (CONE) + NVF (CONE) = 3 x 37 = 111 = ONE CHRIST, A SEED CHRIST of 33 revolutions of Earth around the Sun.

X

Vedic Matheematical Operations

(Grand Design of Nature Imprinted on Sun Rays)

Bended Line	Bended Square
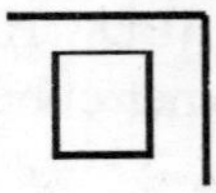	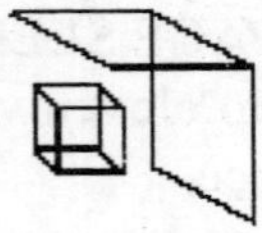
1-space body at boundary of 2-space	2-space body at boundary of 3-space

VMO-1

BEGINNING

(To begin from the beginning)

Ganita Sutra-1 (Ekaadhiken Purvena:/one more than the previous one) accepts organisation for its text as of two parts (first, Ekaadhiken and Second, Purvena). This sutra is the first sutra and it presume the existence of 1 and its rule takes us to '1 more then previous, 1' i.e. to 2. Therefore, the beginning is to be from the artifice of whole number-2 as 1.

NVF of BEGINNING is B=2, E=5, G=7, I=9, N=14, N=14, I=9, N=14, G=7 Total: 81, which is precisely the NVF of SQUARE as S=19, Q=17, U=21, A=1, R=18,. E=5 Total: 81. Therefore, beginning is to be from square, which is a representative regular body of 2-space. Such a beginning is the beginning on the artifice of whole number-2.

VMO-2
JOINT
(The way 'one' joints with 'one')

NVF of JOINT, J=10, 0=15, I=9, N=14 and T=20 is total: 68 = 31+6+31 NVF of CUBE , C=3, U=21, B=2, E=5 Total: 31

Therefore, joint is cube + 6 half damages + cube.

The above help us appreciate the permissibility of accepting this joint as 1+1 as precisely 6 half damages would permit expression as three half damages + three half damages. Therefore, the arrangement 31 + 3 + 3 + 31 = 34 + 34 = NVF ONE + NVF ONE.

The Devnagri script accepts 3 and 6 as pair of reflection images of each other. The format beneath this organization is as of Ganita Sutra-2.

Ganita Sutra-2 (all from nine and last from ten) accept a format for 1 as 01 and settles the range for the image of object '01' through mirror as 10. Therefore, the object, as such remains uptill 1 to 9 and image comes thereafter as last, ten. The image of 01 itself is one entity like the object. Therefore, it is BAG with NVF B=2, A=1 and G=7 Total: 10. We may number it BAG-1.

VMO-3
LANGUAGE / STATEMENTS
(For joints of joints)

NVF of LANGUAGE is L=12, A=1, N=14, G=7, U=21, A=1, G=7, E=5 Total: 68. = NVF of JOINT

NVF of STATEMENTS: is, S=19, T=20, A=1, T=20, E=5, M=13, E=5, N=14, T=20, S=19 Total: 136

136 = 68 + 68

NVF of JOINT, J=10, 0=15, I=9, N=14 and T=20 is total: 68 = 34+34

NVF of ONE: O=15, N=14, E=5 Total = 34 = 17+17

Northern hemi-sphere has the placement value 17.

The NVF of MIDDLE is M=13, I=9, D=4, D=4, L=12, E=5 Total = 47 = 30+17.

The NVF of UNIT: = U=21, N=14, I=9, T=20 Total = 64.

The NVF of ZERO = Z=26, E=5, R=18, O=15, Total =64

64=47+17=30+17+17= 30+34=NVF of "ONE BIBLE" as O=15, N=14, E=5, B=2, I=9, B=2 L=12, E=5 Total : 64 = 4x4x4

With this the joints of joints can be given language/ statements.

VMO-4
MATHEMATICS
(From Square to Cube)

NVF of MATHEMATICS: is M=13, A=1, T=20, H=8, E=5, M=13, A=1, T=20, I=9, C=3, S=19 Total: 112 = 81+31 = NVF of SQUARE +NVF of CUBE.

"Mathematics" has its definition, working domain and formats imbedded in itself in its word composition.

VMO-5
ONENESS IS TO UNDERSTAND AS THAT "ONE GO" IS "ONE"

1. "One Go" is "One".
2. "One" as "One Go" is to express "One Go One".
3. It is a shift from one axis to a pair of axes.
4. It is a shift from one axis to two axes transforming as a pair of axes.
5. It is to join two axes to have a pair of axes.

6. It is to have a joint for two axes as a pair of axes with the help of zero entity as an unit entity in a 4-space domain (43).
7. It is to transform from "bended" line to a "square" with the insertion of MIRROR at the joint. NVF of BENDED is 34 = NVF of ONE, NVF of MIRROR is 91.
8. NVF of ONENESS is 91 = NVF of MIRROR

   O=15, N=14, E=5, N=14, E=5, S=19, S=19
9. The definition of ONENESS is well expressed and is to be comprehended and to have its mathematics by chasing oneness as is of a pair of axes as a bended line with mirror embedded at the joint and the attainment of transformation from linear order of a single axes to a spatial order of a pair of axes.

VMO-6
CONTINUUM AND TRANSCENDENCE

NVF of CONTINUUM is 130

NVF of TRANSCENDENCE is 125

5 SPACE as Origin of 4 space gives us continuum of 4 space as 125 + 5.

This as such gives the grand design of nature for maintaining continuum and transcendence.

It is this grand design which is to be chased as Vedic Mathematical Operations.

VMO-7
SOURCE-SQUARE-BEGINNING

Lord Brahma, the Creator, the supreme, is a four head Lord with each head with a pair of eyes, which manifests creator's world as a spatial order space i.e. a 4-space with 2-space in the role of dimension. This as such settles source as a square as a beginning.

NVF of SOURCE is S=19, O=15, U=21, R=18, C=3, E=5 Total 81

Which is NVF of SQUARE as S=19, Q=17, U=21, A=1, R=18, E=5

Therefore, Beginning is to be with the SOURCE-SQUARE, as

NVF of BEGINNING is also B=2, E=5, G=7, I=9, N=14, N=14, I=9, N=14, G=7

Total: 81

Therefore, the source of Vedic Mathematical Operations is the square and so the beginning is with a square. As such, the forms, formulations and the combination rules for the Alphabet letters scripts, may it be Vedic Alphabet of 52 letters or of English Alphabet letters (26) is to be had from the structural arrangements of the organization format of a square as a representative regular body of 2-space.

VMO-8
1, 2, 3, 8

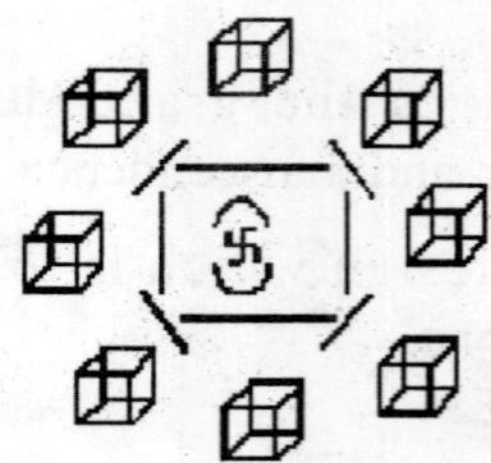

1. Swateshwra-Upanishad enlightens us as that one should process as "1, 2, 3, 8".
2. The artifices of quadruple "1, 2, 3, 8" gives us the organization - arrangements as

   1X1, 2X1, 3X1, 4X2.

3. The Geometric message of above organization - arrangements is that the linear order is uptill 1, 2, 3 and henceforth is spatial order.
4. As such 4-space is a spatial order space.
5. Four spatial dimensions constitute a dimensional frame of 4-sapce.
6. This frame gives us 8 half dimensions and so 4-sapce body accepts 8 boundary components.
7. Interval has 2 end points, square has 4 boundary lines, cube has 6 surfaces and hypercube-4 in a sequence accepts 8 solid boundary components.
8. Hypercube-4 as representative regular body is parallel to the format of Idol of Lord Brahma, the creator, the supreme.
9. Lord Brahma meditates upon his lord, Lord Shiv and multiples as 10 Brahmas. It is parallel to the format of Idol of Lord Shiv.
10. The spatial order of creator's space, as such manifest as MIND LOCK SQUARE as NVF of MIND LOCK = 81 = NVF of SQUARE.

VMO-9
INTERVAL LOCK TO MIND LOCK SQUARE TO CUBE LOCK HYPER

1. NVF of INTERVAL LOCK is 101+41 = 142.
2. 142 = 71 +71 = NVF of SPHERE + NVF of SPHERE.
3. Therefore Interval Lock is of the format of a pair of spheres.
4. NVF of MIND LOCK is 40+41 = 81.
5. Therefore, Mind Lock is of spatial format.
6. CUBE LOCK is HYPER as NVF of CUBE LOCK is 72=NVF of HYPER.

7. As such INTERVAL LOCK to MIND LOCK to HYPER are the orders to be chased as Vedic Mathematical Operations of the Grand Design of the Nature.

8. These operations are to reach at format of CREATOR, as MIND LINE: to pullout Creators world out of continuum, to workout circumference, to half the continuum to be at center etc. etc.

VMO-10
ONE, FREE; LOCK, KEY

For the facility of expression, I am henceforth accepting a mode of writing the technical words in capital letters with its NVF in the braces following, for example, ONE (34), FREE (34), LOCK (41), KEY (41).

1. Origin, of 4-space is 5-space.

2. 5-space is a solid dimensional space.

3. 3-space has three linear dimensions emerging from a common origin.

4. Three dimensions of 3-space give us 6 half-dimensions emerging from the origin, which is of spatial order of 4-space.

5. Six half-dimensions and seventh origin, together as seven entities are the basic organization of dimensional frame of 3-space.

6. It is because of the spatial order of the origin, which permits a pair of half-dimensions for each dimension of 3-space and the spatial order organizes them as two sets of three "half-dimensions" each, with each set on the either side of the origin at the middle.

7. The placements of two sets of three "half-dimensions" each, on either side of the origin at the middle, are the placements at two different place values, the first set of three half-dimensions, as three entities, at a unit place value and other

set of three half-dimensions, as three entities, at ten place value, with total value as 30. This makes the value as 33.

8. The spatial order, which takes to (43) with potentiality of ZERO(64) transforming as UNIT(64), makes the above arrangement of two sets of half-dimensions as of value 33.
9. Further, the manifestation of 4-space format as hypercubes, and hypercube-4 as a representative regular body of 4-space gives us domain-boundary ratio for it as A4:8B3 as the boundary of hypercube-4 is constituted by eight solid components (each as a cube).
10. Therefore, the eight boundary components lock the domain and this as such gives us 8 (entities due to boundary) and 33 (value of the entities of the arrangement of the two sets of half-dimensions) together as 8 + 33 = 41 = LOCK (41) = KEY (41)).

VMO-11
SUNLIGHT TRADITION

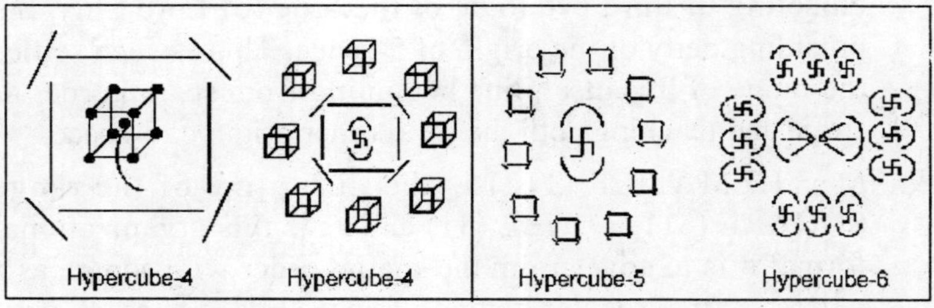

1. Tradition goes as that the Vedic knowledge is organized on sunlight parallel to the grand design of nature imprinted on sun rays.
2. SUNLIGHT(110) is of NVF 110 and so is TRADITION (110).
3. Jyoti flows from core of the sun on its rays as sunlight and it fills the manifested boundary of Sanatana (5-space) and creates CONTINUUM (130).

4. **Sun as a body** of 6-space, admits 13 geometries and the 10 boundary components of hypercube-5 that way get fulfilled with 13×10=130 entities of the order of 6-space.
5. As each of the boundary component of hypercube-5 is hypercube-4, so the same (hypercube-4 domain) gets fixed in terms of a 13-edged cube.
6. With 5-space as origin of 4-space, the continuum order of 130 entities after separating five entities thereof for the origin, makes available 125 entities which are of the order admitting organizational arrangement as 53.
7. The availability of organizational arrangements of 53 within and around 5-space as origin is to be chased to work out the grand design of nature imprinted on sun rays as is being chased as Vedic mathematics, science and technology as a discipline of knowledge preserved in and enlightened as Sthapatya Upved.
8. Manasara is the scripture of Sthapatya Upved which begins with the prayer of Lord Brahma, the overlord of 4-space and the enlightenment of Sthapatya Upved completes with chiseling of third eye to be of the order of Lord Shiv, the presiding deity of the origin of 5-space. This as such settles the range of this discipline beginning from spatial order 4-space and reaching uptil the solid order origin of 5-space.
9. MATHEMATICS (112) is the discipline of the range SQUARE (81) to CUBE (31) and as is this organizational format it is to cover from the spatial order with square as a representative regular body of 2-space and is to reach uptil solid order of the origin with a cube as a representative regular body of 3-space.
10. The CONTINUUM (130) with 8 entities for the boundary and 5 entities for the origin makes available 117 entities, of which when 5 are reserved for the seat of the origin, because of its dual role, as a part of the 4-space domain, as well as a part of the 5-space domain, we have the availability for 4-space domain, with origin sealed (that is by excluding the role for the origin point of 4-space as that of 5-space as seat of origin), of 117-5=112 entities.

11. To pull out of the continuum the domain of 4-space, we have to pull out 13-edged cube in terms of which the 4-space domain stands fixed in the continuum. This operation to pullout 4-space domain out of continuum in terms of 13-edged cube is the operation of the grand design of nature imprinted on sun rays. With pulling out of 4-space domain, we would be left behind origin of 4-space i.e. 5-space (of 5 entities) and the boundary of hypercube-4 i.e. 8 solid components (of 8 entities). As such the pulling out operation of Nature on the rays of sun is the operation of 130-13=117 entities.
12. The 13-edged cube is a cube of 13 edges and 9 points, of which 8 remains to be the edges of the cube and ninth is the center of the cube. This as such gives us the organizational arrangement for the artifice of 117 as 13×9.
13. To PULL (61) OUT (56) means to have CHURCH (61) OUT (56). As such this pulling out operation of domain of 4-space from the continuum is the operation to have a CHURCH out of the continuum.
14. The organizational arrangement of OUT (56) is there because of the organization arrangement permissible at the boundary of hypercube-4 consisting of 8 solid components / cubes, as the cube has a structural set up as much as that it permits connecting its all the eight corner point in terms of precisely seven edges only and this for all the eight boundary components give us 7 × 8 = 56.
15. The property of spatial order to provide di-monad format when worked in reference to MIDDLE POINT, it synthesizes object and image as a reflection pair which makes the seat of origin as middle of the domain. As such MIDDLE (47) and POINT(74), together give us 47 + 74 = 121 = 11 × 11 which is organizational arrangement for the seat of origin of 4-space. With 5-space in the role of origin of 4-space and with 5-space admitting precisely 11 geometries will help us appreciate the organizational format of 11 × 11 as the organizational format for the middle point

of 4-space domain. This further gives us an insight as that 125–121 = 4 is parallel to the artifice of 4 as of the order of 4-space.

16. The dimensional order of 5-space as $1 \times 3 \times 5 = 15$ would further help us have an insight as that 125–15=110 is of the order of SUNLIGHT (110). Further as the domain-boundary ratio of hypercube-5 is A5:10B4, therefore, 10 entities for the 10 boundary components of 5-space will help us have the balance as of 110–10 = 100 entities.
17. Still further by preserving 56 entities reserving for OUT / 8 boundary components of hypercube-4, we would be left with the balance 100–56 = 44.
18. This balance 44 is precisely the NVF of RAY, SPACE, BRAIN, FAITH, HOPE, HIDDEN, PAIR.

Exercise Chase:

1. Draw split ups of $125 = x + y$ as parallel arrays of words of corresponding artifices as x and y.
2. One split up of artifice of 125 is 47+78. This split up is availed in the organization format of Srimad Bhagwad Gita as much as that the first chapter of Srimad Bhagwad Gita is of 47 Shalokas and the last i.e. the 18th chapter of Srimad Bhagwad Gita is of 78 Shalokas.
3. Artifice of 47 is the artifice of MIDDLE. One may enlist the arrays $125 = x + y$, $x = 47$, $y = 78$. The words ADMIT, CURE, EARNED, AUDITED, FEELS, FRAMED, MIDDLE, MONAD, NEEDS, RIGID, SAVE, TIME accept 47 as NVF. One may draw and extend this array of words of NVF 47. Likewise one may draw an array of artifice of NVF 78. Then one should try to see how these split ups $125 = x + y$ are the different aspects of the domain of 4-space.

VMO-12
FOUR CUBES

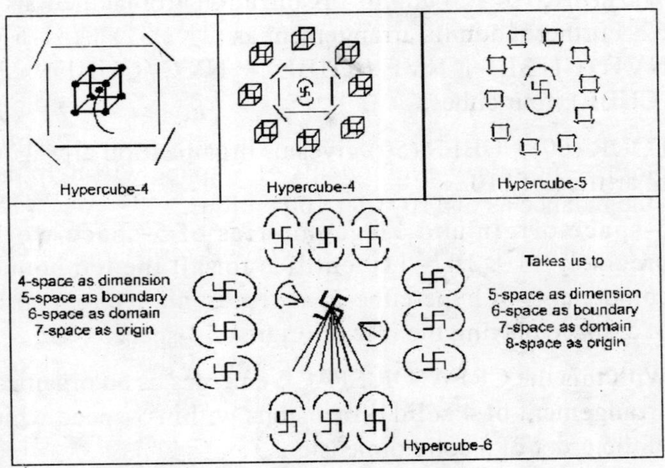

FOUR CUBES

1. CREATOR(80) SPACE(44) gives us organization arrangement of 80+44 = 124.
2. The artifice of 124 has powerful organization arrangement as:

1	2	3
2^0	2^1	2^2

3. The artifice of 124 also admits organization arrangement as 123+1.
4. CONTINUUM (130) can be exhaustively covered as 125 + 5, where 5 is reserved for the origin seat. The artifice of 125 gives us arrangement as 1 + 123 + 1. This is an arrangement of exhaustive coverage of the continuum domain with origin sealed as the domain outside circumference, circumference and within circumference. This is when given one entity for outside circumference and one entity for inside circumference, shall give us 123 entities for the CIRCUMFERENCE (123).

5. CREATOR SPACE (124) is the domain within circumference including circumference.
6. The artifice of 124 admits organization arrangement as 62 + 62. Further it admits arrangement as (31 + 31) + (31 + 31) = NVF (CUBE) + NVF (CUBE) + NVF (CUBE) + NVF (CUBE); four cubes.
7. FOUR (60) CUBES (50) gives us organization arrangement of artifice of 110.
8. 5-space origin and 11 geometries of 5-space workout precisely 11 × 10 = 110 entities for all the ten boundary components of hypercube-5, as representative regular body of 5-space playing the role of origin of 4-space.
9. With this the CREATOR SPACE emerges as an organization arrangement of 4 solid dimensions within 5-space which is of the order of 5 solid dimensions.
10. With 1 × 3 × 5 = 15 as dimensional order of 5-space, and 5 entities for the seat of origin within 4-space, reserve remaining 130 – 15 – 5 = 110 entities of continuum for the creator space with potentiality to exhaustively cover as boundary of hypercube-5.
11. CONTINUUM (130) organizes itself as CREATION (105) AREA (25). The CREATION (105) emerges as of the order of 105 = 1 × 3 × 5 × 7 which is precisely the dimensional order of 7-space with 5-space in the role of its dimension.
12. CREATOR (80) gives us the organization arrangement 80 = 10 × 8 which is equal to 80 cubes which cover the boundaries of all the ten hypercube-4 boundary components of hypercube-5.

VMO-13
FOUR SQUARES AND FOUR MIDDLE RAYS

1. Four cubes work out 4-space within four solid dimensions in 5-space.

2. Four squares work out as 4-space.
3. FOUR (60) SQUARES (100) as 160 NVF is of the order of NVF (MIND) + NVF (MIND) + NVF (MIND) + NVF (MIND).
4. NVF of SQUARE is 81. As such 4 × 81 = 324 is an arrangement which further permits expressions as 18 × 18.
5. The expression of NVF (SUN) = 54 = 17 + 18 + 19 takes us to 17 in northern hemisphere, 18 at middle and 19 in southern hemisphere.
6. NVF (MIDDLE) = 47 and the NVF (RAY) (44), gives us 47 + 44 = 91.
7. 91 NVF is of MIRROR.
8. As such the organization of spatial dimensional order at the middle shall be taking us to the organizational shift from that of FOUR SQUARES to that of FOUR MIDDLE RAYS.
9. This shift from that of FOUR SQUARES to that of FOUR MIDDLE RAYS is in fact a shift from 4-space manifestation to that of transcendence at the origin of 4-space into 5-space.
10. This is an organization which takes us from 4-space domain to its origin (5-space) and it is the phenomenon of SUN RISING (130) which is of the order of CONTINUUM (130) and the WHITE (65) is half of the CONTINUUM. It is like one phase of monad being white and the back of the monad as black. WHITE (65) with NESS (57) plus 8 entities for the boundary components of hypercube-4 would give us the continuum of hypercube-4 including that of its boundary. It is this WHITE CONTINUUM which merges boundary with the domain and it is this phenomenon of manifestation which proves as a cover which need be removed to have the transcendence.

VMO-14
BOTTOM END TO WORLD TOP

The number within parentheses ( ) is NVF of the word preceding it. And, the number within Square Brackets [ ] is sum of NVFs of the words preceding it.

1. TOP (51) is FULL (51).
2. WORLD (72) is HYPER (72).
3. SUN (54) is CUBE END [54].
4. MATTER (77) is SUN END [77].
5. CONTINUUM (130) is that bottom end to go; BOTTOM END GO [130].
6. WORLD TOP [123] is the circumference of the domain enclosed.
7. The HYPER (72) WORLD (72) gives us NVF 144 which admits organization arrangement as 12 × 12.
8. The three space in the role of dimension takes us to 6-space as origin.
9. The domain-boundary ratio of hypercube-6 comes to be A^6:$12B^5$.
10. Therefore, the orbitals 1 × 2, 2 × 4, 3 × 6, 4 × 8, 5 × 10, 6 × 12 ultimately takes us to hyper world.
11. This from bottom end to world top is the hyper world continuum with Sun as the origin of the whole range of matter.

VMO-15
THIS (EYE) TO THAT (EYE)

1. Creator the Supreme is a four head Lord with two eyes in its each head.
2. Creator the Supreme meditates upon His Lord, Lord Shiv, the Lord of five heads with three eyes in each head and multiplies Himself as ten Brahmas.

3. Chiseling of third eye, the creator's way is to go from "This Eye of Creator" to

   That Eye of His Lord, Lord Shiv".

4. THIS (56) is of NVF order of CHISEL (56).

5. THIS (56), THAT (49) together give us $56 + 49 = 105 = 3 \times 35 = 3 \times$ NVF of EYE (35).

6. THIS EYE TO THAT EYE (210) admits organization of NVFs of 6 eyes as $6 \times 35 = 210$.

7. WHITE (65) CONTINUUM (130) to BLACK BACKS [65] CONTINUUM (130) is a range which takes us from one part of di-monad to second part of di-monad of the spatial order of Creator's space. It is this which takes us from the spatial symmetry to solid symmetry with chiseling at the origin of 2-space / center of a square.

8. With seat of 5-space at origin of the continuum domain of 4-space, the origin / center accepts 5 entities for its expression and these five entities together with NVF of SQUARE (81) gives us $81 + 5 = 86$, the NVF of CHISELING (86).

9. "This (Eye) to That (Eye)" is the phenomenon of chiseling of third eye by chiseling at the origin of spatial order.

10. The chiseling of the spatial order would take us to the solid order and with it our study zone would shift to solid dimensional order of real 5-space with 6-space in the role of origin.

11. This settles study zone as 6-space as origin - domain and 5-space as domain-boundary.

12. The availability of the solid dimensional order permits the split up for the sphere as northern and southern hemispheres and the continuum emerging in between.

13. The availability of solid-dimensional order further permits to go to middle of middle. It is like going to the sphere end as NVF (MIDDLE - MIDDLE) = 94 = NVF (SPHERE END).

14. FRONT of monad and BACK of monad takes us to NVF (FRONT - BACK) = 90 which admits arrangement as $18 \times 5 = 90$.

15. The six surfaces of the cube permit us to strip off the solid content of cube by removing the solid content along six surfaces with each surface moving outward would be creating three dimensional order and with this we shall be having $3 \times 6 = 18$ dimensions order for removal of the solid content of cube.

16. This as such works out organization arrangement of $90 = 18 \times 5$ as an arrangement of 5 solid dimensions, which is the arrangement of the solid dimensional order of 5-space.

17. With this our start with spatial order of 4-space as **creator** space takes us to the solid order of 5-space as a space **of** Lord of Creator, the way the creator the supreme meditates in the cavity of His Heart upon His Lord, Lord Shiv **and** multiplies Himself as ten Brahmas.

18. This is the way the knowledge range of Sthapatya Ved has been arranged starting with the prayer of Lord Brahma and reaching uptil the stage of chiseling of third eye.

VMO-16
ENTITY - ENTITIES

1. **ENTITY** (93) admits organization arrangement as 3×31 and as 31 is NVF of CUBE, which is representative regular body of 3-space and hence it takes us to the domain within 3 solid dimensions. Therefore, entity is the geometric entity of the domain within three solid dimensions.

2. ENTITIES (101) have NVF equal to that of an INTERVAL (101). If we look at the expression of an interval on a dimonad format with 5 entities of which 2 parts of interval are fixed in terms of three points. As such, the geometric arrangement here turns out to be that of a solid dimensional order that is of real 5-space within 5 solid dimensions set up.

3. As 1-space is in the role of dimension of 3-space and 3-space is in the role of dimension of 5-space, as such, the above arrangements and comprehensions for entity and

entities give interesting insight into the way the compositions of di-monad format work out a mathematics.

4. It is this mathematics, whose study zone is being searched as the domain of Sun. SEARCH (54) is of the order of NVF of SUN (54).

VMO-17
CONTENT MIRROR AND MIRROR CONTENT

1. Content mirror and mirror content are two different aspects. The content mirror is the aspect of content having potentiality to organize and workout as a mirror. The mirror content is the aspect of the mirror as content.

2. The potentiality of CONTENT (91) as of MIRROR (91) is there because of their formats accepting common NVF.

3. The artifice of 91 admits arrangement as 13×7 which speaks for its potentiality as each of the 13 edges of 13-edged cube (hypercube-4) getting optimum linear range as of permissible seven steps parallel to the potentiality of 7 edges to connect all the eight corners. So would get connected all the eight boundary components of hypercube-4 in seven steps. This as such becomes the potentiality of the artifice of whole number 91 to have full expression for all the 13 edges of 13-edged cube within 4-space. It is this potentiality of artifice of whole number 91 to simultaneously workout linear and spatial orders which interconnects CONTENT and MIRROR.

4. Mirror extends the linear order to spatial order by interconnecting object with the image.

5. The mirror operation of connecting object with the image and transforming monad into di-monad is the operation of locking the interval on a di-monad format.

6. INTERVAL LOCK [142] is of NVF of the order of a pair of spheres as $142 = 71 + 71$ = NVF (SPHERE) + NVF (SPHERE).

7. Two spheres lock interval and this may be chased as a depiction as under:

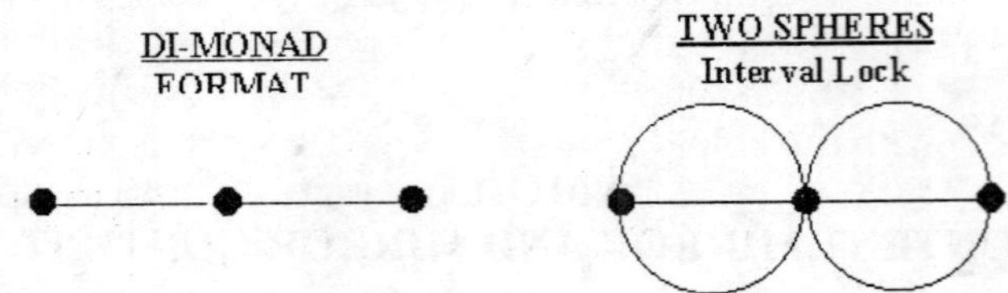

8. The two parts of di-monad have their individual identities because of the middle joint. The middle joint as one entity, would when removed take us from 101 NVF of INTERVAL to that of 101–1 = 100. With the removal of the joint, we shall be left with two parts, each of which would be an interval of monad format. Each part on a di-monad format would again take us to the monad format interval and such part of the part of di-monad interval would in fact be one quarter of the original interval of di-monad format. This would help us have an insight as two way NVF (QUARTER) is 100.

9. NVF (QUARTER) as 100 is equal to NVF (SQUARES). This as such gives us an insight as that the squares are to be organizationally approached as quarters. This further would give us an insight as how the spatial order is to be worked out as four spatial dimensions.

10. For the further insight of working out squares as quarters as spatial order of four spatial dimensions, firstly we have to see the organization arrangement of the artifice of whole number 100 as 10×10. This would take us to the organization arrangement of whole number 10 = 1 + 2 + 3 + 4. This further would take us to the domain-boundary ratio of hypercube-5 as A5:10B4. With this, we may have an insight as to how 10×4 = 40 coordinates requirement for exhaustively coverage of the boundary of hypercube-5 would give us a connecting bended Line with NVF (LINE) as 40 and NVF (BENDED) as 34 = NVF (ONE).

11. Bended line coordination of all the 40 coordinates for all the 10 boundary components of hypercube-5, as such is the linear order on spatial format and with this order at the boundary of hypercube-5, we have to reach within the solid format of domain of hypercube-5 in terms of the di-monad format for a pair of squares / planes / 2-space. This would be possible with the help of insight to ADD ALL to get ONE as NVF (ADD ALL) = NVF (ADD AREA) = 34 = NVF (ONE).

12. The dimensional order of 5-space is solid order. Its linear equivalence is 1 × 3 × 5. The spatial order for 1, 3, 5 shall be 12, 32, 52. The artifices of 91 with 32 at ten place value and 12 at one place value and further 34 = 9 + 25 = 32 + 52 are the organization arrangement availing 12, 32, 52.

13. The organization format for Brahm Rishi Ved Vyas, who organized single Ved as four Veds, is the format of Vyas (Diameter). And this would help us have insight into the organization format of Veds and Vedic knowledge.

14. The addition and removal of a single point provides transformation from monad format to that of di-monad format for an interval and with it there amounts to a shift from a whole (plane) to that of a quarter (of plane). This factor i.e. 4 as a multiplier is the factor which deserves to be chased to have insight for working with a shift from linear format to spatial format.

15. This would help us have an insight as to how the sole syllable OM is accepted as AUM of four quarters.

VMO-18

LINE LOCK AND T-20

1. Line, as linear order, and hence of 1-space format, is locked and blocked by spatial order and hence in 2-space format with 1-space permitted to remain only at boundary of 2-space.

2. LINE LOCK [81] as such is of the order of SQUARE (81).

3. LINE (40) and LOCK (41) give us an insight as that lock is to be one entity ahead of the entity being locked.
4. KEY (41) as such gives us an insight that entity locked when unlocked with KEY is to take us to a domain of one entity less than that of NVF of KEY.
5. LINE (40), as four quarters with each quarter as a block of 10 entities would take us to 10 + 10 = 20 entities for a pair of quarters and as such T as a pair of quarters of a square / plane / 2-space will be of the order of 20 entities and as such NVF (T) = 20.

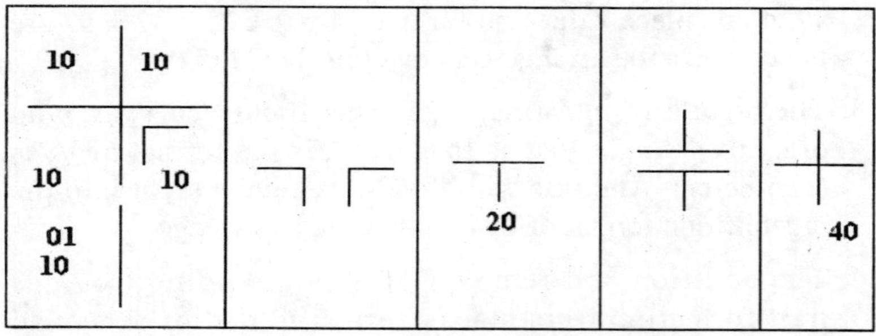

6. LINEAR (59) and SPATIAL (39 + 39) give us the organization format for 59 = 60 – 1 or 59 + 1 = 60 = 30 + 30 and for 39 + 39 = (40 – 1) + (40 – 1) or 40 + 40 = (39 + 1) + (39 + 1). And would help us have an insight of how the lock as one entity ahead of entities locked and key unlocking and taking us one entity less than the locked entities. This also would help us have an insight how far linear as an order of 1-space, the locking-unlocking process takes place once and for spatial as an order of 2-space, the locking and unlocking process is to take place twice.
7. The MANIFESTATION (146 = 73 + 0 + 73) and UN-MANIFESTATION (181 = 90 + 1 + 90) would help us have how the same entity is ZERO (64) and is also UNIT (64) because of their placements. The entity at the boundary is zero and the entity in the domain is one. It is this

manifestation and un-manifestation processes which are there in the processes of lock and key.

8. With single entity removed or added takes us from one to quarter and quarter to one and the manifestation and un-manifestation process makes the same entity to be zero and also to be one. It is their because of the spatial order. It is this order whose insight is to help us have a transition from linear order of monad to a spatial order of a di-monad.

9. The study zone, at dimensional level, means a solid dimensional order emanating from hyper solid (HS-4) origin. On the artifices of whole numbers the same may be chased as 34, 4 at a unit place and 3 at a ten value place. With 4 at the middle (origin) and it taking to either side to 3. It is to be of di-monad format with one part as 34. It is NVF of ONE.

10. The reflection image of 34 is 43. Firstly, 34 + 43 = 77. It is NVF (CHRIST).

11. Secondly, the artifice of 77 gives us 7 at a unit place as well as 7 at a ten value place. It is its own reflection image as the replacement of place values does not affect the value of the artifice.

12. Thirdly, the artifice of 7 is unique in its potentialities being the biggest prime numeral of ten place value system and secondly it is of the order of the 7 geometries of 3-space and thirdly it is the optimum of the linear order as much as that 7 edges are capable of connecting all the 8 edges of the cube and it is only uptil hyper circle-7 that we have an increase and from hyper circle-8 onwards go on decreasing.

13. The affine order to be of same value at each place value, as is in case of 77, gives us the freedom of handling the artifice of 7 as a package and as such we may have 7 + 7 = 14.

14. This would give us an insight of the organizational arrangement as well as the format beneath the Gantia Sutras as per which the first Sutra manifests the mathematics of "one more than the previous one" and the fourteenth Ganita Sutra manifests the mathematics of "one less than the

previous one"; the processes together being parallel to the processes of manifestation and unmanifestation as well as parallel to the processes of lock and key.

15. The domain-boundary ratio for the hypercube-7 comes to be A7:14B6, which in other words would mean that there would be a requirement of as many as 14 boundary components of the order of hypercube-6 to complete the boundary of hypercube-7.
16. There are precisely 14 Maheshwara Sutras which coordinate precisely 7 × 6 = 42 alphabet letters. This would give us an insight the coordination of 6 dimensions of 6-space as origin and 7 dimensions of 7-space as origin of origin or the transcendence domain.
17. The origin of the domain has a dual status, firstly as an entity of the domain and secondly as an entity of the origin. It is the unique locking system of locking every domain. This is to go from domain to domain as a lock to lock.
18. LOCK LOCK [41 + 41 = 82] is precisely the NVF of MEASURE.

VMO-19
MIND'S EYE, MINDS & EYES

1. "Lock is to be one entity ahead of the entity being locked" as a Vedic mathematical operation, and a working rule, is a very powerful operation having very rich mathematics of great potentiality. And, the first Ganita Sutra with the working rule "one more than the previous one", is in fact is applied value of the above operation and rule for which an insight may be had with mind's eye.
2. To lock "linear lock" we have to in terms of above operation, one entity more than that of NVF of LINEAR LOCK.
3. One entity more than NVF of LINEAR LOCK [100] takes us to 100 + 1 = 101 which is precisely the NVF of INTERVAL (101).

4. Interval as locked linear-lock is Vedic Mathematical Comprehension (VMC) and result (Vedic Mathematical Result, in short VMR) which would give us an insight of the geometric organization of linear order to that of interval of di-monad format.

5. This VMC and VMR would further give us an insight as to the way MIND (40) which unlocks the lock of KEY (41) and gets unlocked as MIND (40) as of the order, equal to NVF of LINE (40).

6. This would further give us an insight the way MIND (40) arranges and chases MINDS (59) parallel to the LINE (40) chases LINEAR (59).

7. EYES (54) are of the order equal to NVF (SUN).

8. MIND'S EYE [94] is of order of NVF (MIDDLE-MIDDLE) = 94.

9. DI-MONAD (60) is one entity ahead of LINEAR (59); and as such DI-MONAD locks LINEAR (59).

10. DI-MONAD (60) locks LINEAR (59); and this becomes Vedic VMC and also VMR, which may be designated as an ORDER (60).

11. DI-MONAD (60) as an ORDER (60) deserves to be comprehended and chased with mind's eye.

12. The above locking and unlocking processes of Vedic Mathematical Operations because of the simultaneous permissibility for the ZERO (64) and UNIT (64) as of equal NVFs would permit us handle simultaneously LOCK (41)-KEY (41); LINEAR (59)-ORDER (60); DI-MONAD (60)-ORDER (60).

13. This also would help us have an insight as to that for transition from 10 × 4 = 40 coordinates fixation of the boundary of hypercube-5 for which domain-boundary ratio is A5:10B4 to that of 12 × 5 = 60 coordinates fixation of the boundary of hypercube-6 for which domain-boundary ratio is A6:12B5, we have to transit from MIND (40) to ORDER (60) parallel to transition from LINE (40) to DI-MONAD (60).

14. As such, it may be concluded, as that, within study zone, by operating from boundary to boundary, we may have the desired attainment of transiting from solid dimensional order to hypersolid-4 dimensional order.

VMO-20
BEND AREA, BENDED MIDDLE

1. To bend and get area, is a line management operation for which, insight may be had as that with a bend, the line gets the potentiality to have an expression along a pair of axes and with it is got area within the pair of axes.

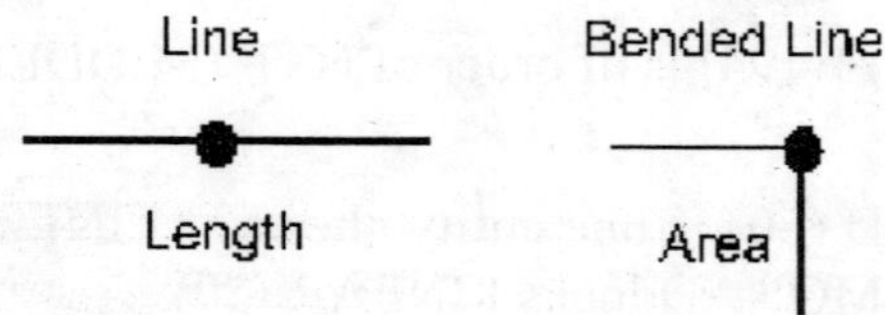

2. BEND (25) and AREA (25) would further help us have comprehension of this management aspect.
3. To bend area, parallel to the line management operation, would be a plane management operation. A bended plane likewise would have an expression along a pair of spatial axes. This straight-a-way would give us AREA (25) along the first axis and AREA (25) along other axis.
4. To bend (AREA) and get (AREA) is a management operation for whose insight we may have by going to NVFs of BEND (25), AREA (25) and AND GET [50].
5. This would further help us comprehend as that the bended plane, as an expression along a pair of spatial axes, would be enveloping a "void".
6. VOID (50) would help us have an insight with organization arrangement for the artifice of 50 as 25 + 25; void enveloped by a bended plane.

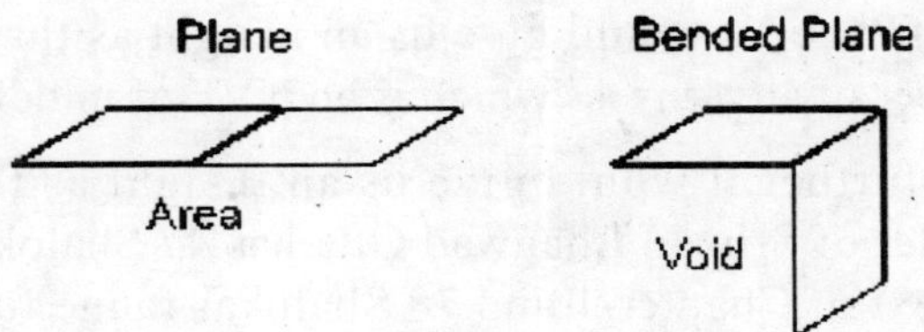

7. Middle of a di-monad format has linear order along its first part and a spatial order along its second part because of the bending potentiality at the middle. This frames monad within a plane. The domain-boundary ratio of 2-space body as A2:4B1 requires four boundary components. With this, four entities for the boundary would take us to NVF (MIDDLE) = 47 - 4 entities for boundary = 43 as an artifice for the FRAME (43) of the monad.

8. The bending potentiality of the middle of a di-monad, as an aspect accepted as bended middle, will give us an insight as management of the second part of the di-monad i.e. of the range from middle to the second part i.e. from middle to spatial part i.e. MIDDLE-SPATIAL.

9. MIDDLE (47) and SPATIAL (78) would give us an insight with organization arrangement for the artifice of 53 = 125 = 47 + 78.

10. We may have further insight in this management operation with the help of the organization arrangements of the artifices of numbers 47 and 78 as that 78 = 47 + 31 = NVF (MIDDLE) + NVF (CUBE).

11. The artifice of 47 admits organization arrangement as 47 = 16 + 31 and further the organization arrangement for the artifice of whole number 16 as 5+6+5 takes us to the dimonad format for the study zone with 6-space at the middle and 5-space at either side.

12. This would further give us an insight by going to the NVF (BENDED) = NVF (ONE) = 34 and NVF (MIDDLE) = 47 and together NVF (BENDED-MIDDLE) = 81 = NVF (SQUARE).

13. Still further it would give us an insight by going to NVF (SQUARE) = 81 = NVF (SILVER).

14. Still further it would give us an insight as that the atomic number of silver is 47 which is an NVF of middle.

15. Still further it would give us an insight as that the first chapter of Srimad Bhagwad Gita has 47 Shalokas range for the text of Chapter-1 and 78 Shalokas range for the text of its last Chapter i.e. Chapter-18.

16. The "middle's frame" as an artifice of whole number 43 can be chased as 17 + 26 = 1 + 16 + 25 + 1; 1 + 16 taking us to locked artifice of 16 and 25 + 1 taking us to the locked artifice of 25.

17. The artifice of 25 is NVF of AREA; the artifice of 16 is the total number of arrangement for an interval as of three parts. Therefore, the artifice of 43 = (1 + 16) + (25 + 1) is the organization arrangement of locked artifice of 16 as of trimonad and the locked artifice of 25 as of dimonad.

18. This would give us an insight for the middle as permitting management in terms of simultaneous availability of tri-monad and di-monad.

19. This would still further give us an insight for the middle permitting management in terms of simultaneous availability of cube and a square.

20. This as such would give us an insight that mathematics is the mathematics of middle as bended middle.

VMO-21
SUN GOD

1. SUN GOD [54 + 26 = 80] is CREATOR (80).

2. SOUL (67) goes to RAY END [44 + 23 = 67].

3. SOLE SOUL [51 + 67 = 118] is of the order of NVF (LINEAR) + NVF (LINEAR) = 59 + 59 = 118.

VMO-22
TRINITY OF GODS

1. TRINITY (115) OF (21) GODS (45).
2. TRINITY is of 115 as NVF of the order of artifice 115 which admits organization arrangement as 5 × 23. The NVF (END) being 23, therefore it would be an organization arrangement of 5 ends.
3. OF is of 21 as NVF of the order of the artifice of 21 which admits organization arrangement as 1 + 2 + 3 + 4 + 5 + 6, the measuring-rod which may take us uptil 6-space.
4. GODS is of 45 as NVF of the order of the artifice of 45 which admits organization arrangement as 1 + 2 + 3 + 4 + 5 + 6 + 7 + 8 + 9, the measuring-rod which may take us uptil 9-space.
5. TRINITY OF GODS is of 181 as NVF of the order of the artifice of 181 which admits arrangement as 100 + 81 parallel to the NVF of SQUARES plus NVF SQUARE.
6. The square of squares and squares' square are aspects which when chase would give us insight of the artifice of 181 which also accepts expression in 3 parts with first, unit place value part as 1 × 1 × 1 and middle ten place value part as 2 × 2 × 2 and the other end part also as 1 × 1 × 1.
7. The above expression of 181 as of three parts parallel to three digits permitting expression as 1 × 1 × 1, 2 × 2 × 2 and 1 × 1 × 1 would also give us an insight as the three dimensions of 3-space accepting a tri-monad format as 1, 2 and 1 as expression for three parts of the dimension.
8. The expression 121 for the above format for the dimension would further help us fold and re-organize it as 11×11. This as such takes us to a di-monad format i.e. format of two parts with each part further as of two sub parts with each sub part as 1.
9. This would also give us an insight of transition from a dimoand format to tetra-monad format on the format of a tri-monad format.

10. The artifice of 11 otherwise has the potentialities parallel to 11 geometries of 5-space, a solid dimensional space.

11. NVF of IDOL as 40 would give us an insight as that it has the potentiality to coordinate 10 hypercubes-4 boundary components as a LINE (40) connecting all the 40 coordinates of all the 10 boundary components of hypercube-4.

12. The spatial order of the boundary components of hypercube-4 would further help us have an insight as to how the dimonad format for the linear order would double the NVF for the LINE (40) as NVF (LINE) + NVF (LINE) because of dimonad format and with it the attainment of 80 as NVF shall be making it potentially of the order of NVF of CREATOR (80) as NVF of SUN GOD [80].

VMO-23
HYPER SURFACE MANIFEST FROM UN-MANIFEST

1. Hyper surface is responsible for transformation of a cube into a pair of prisms.
2. NVF (HYPER SURFACE) = 62 is of the order of NVF (CUBE) + NVF (CUBE).
3. This organization permissibility for NVF (HYPER SURFACE) as NVF of 2 cubes is there because of the acceptance of a di-monad format for an interval as a linear dimension.

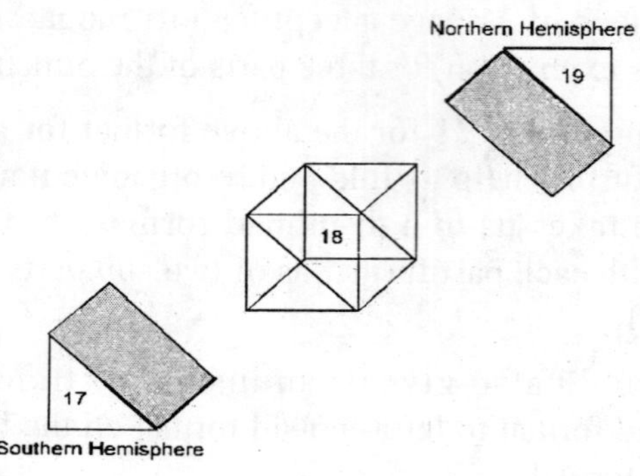

4. The un-manifest joint of di-monad manifest because of the spatial order of di-monad having the potentiality to act as spatial dimension to create real 4-space of spatial order and within real 4-space, the simultaneous permissibility of manifesting as linear order as well as spatial order makes out a situation where NVF (ZERO) = 64 = NVF (UNIT).

   This is there as within real 4-space, the dimensional bodies manifest as of 4 folds with n-space playing the role of boundary of n+1 space. It is because of it that the same entity while at boundary is of zero value domain and the same otherwise is unit entity of the space playing the role of boundary.

5. Of the four folds, of manifestation within real 4-space, the origin fold gets sealed and with it the manifest and un-manifest gets separated.

6. The four folds of manifestation, with each fold itself as a manifested fold are expressible in Vedic language as Vyakta, Avyakta, Avyakto-Avyaktat and Purusha. In terms of artifices of whole numbers these may be expressed as the artifices of 3, 4, 5 and 6 respectively.

   These together sum up as 3 + 4 + 5 + 6 = 18. This is precisely the location value of the hyper surface within a solid which on separation of its two parts, first with location as a northern hemisphere and second with location as southern hemisphere get association of the number values as 17, 18 and 19 respectively and these together sum up 54 as of the order of NVF (SUN).

7. The four folds of manifestation are separated by the hyper surface as un-manifest Trinity of Gods from Purusha (human body frame). This is like the separation of the manifestation of human frame with whole manifestation on one side and same with its back ends, gets sealed, locked and limited.

8. UN MANIFEST (181) TRINITY OF GODS [181] gets separated from manifest Purusha as BACK END SEAL LOCK LIMIT [17 + 23 + 37 + 41 + 63 = 181].

9. This takes us to 181 + 181 + 181 = 543; which is the organization arrangement of the order of tri-monad with its three parts sequentially increasing for the values at the place values parallel to the increasing order of the place values format along the three parts of the tri-monad format. This achievement is the achievement for the opposite orientation for the order and organizational arrangement of 1, 2 and 3 as 123, which otherwise is the orientation of circumference.

   This that way comes to be an enlightenment that while going along one side orientation along a line, as +1 space we organize and arrange and manage by availing the geometric format of +1 space in the role of a domain fold. However, while we avail the reverse orientation along a line as (-1) space. What is being availed is the geometric format of 1-space in the role of dimension.

   With a shift from +1 space orientation of a line to (-1) space orientation of a line is the difference of a shift from a domain fold to dimension fold. This difference in fact is the difference of moving from n-space in the role of domain fold to that of (n–2) space in the role of a dimension fold.

   The NVF 123 of a CIRCUMFERENCE is as number value an organization arrangement of the orientation of a decreasing order as here 3 is at a unit place value, 2 is at ten place value and 10 is at hundred place value.

   Therefore this is of 1-space geometric format as a dimension fold. For a shift from dimension fold to the domain-fold we have to a shift of the order (n–2) to n. This as such gives us a rule of increase of 2 with whose application we get the numbers 1, 2 and 3 of increased values as 3, 4 and 5. Then the orientation placements for them would make it as a number with 5 at a unit place value, 4 at a ten place value, and 3 and hundred place value and the number yield would be 543.

10. The affine order is free of orientations. This works out as a difference of orientation values 543 and 123 as 420. One of the organization arrangements for 420 is 42 × 10 = (21 + 21) × 10. As such this arrangement gives us artifice of 42 for

each of the 10 boundary components of hypercube-5. The artifice of 42 as 21 + 21 is parallel to the artifice of a measuring-rod as 1 + 2 + 3 + 4 + 5 + 6. Therefore, the artifice of 21 + 21 gives us a measuring-rod of a di-monad format such that each of the two parts being a measuring-rod 1 + 2 + 3 + 4 + 5 + 6.

This availability of a synthetic measuring-rod as of two parts is parallel to the potentiality of the spatial order of 4-space whose representative regular body hypercube-4 is a boundary component of hypercube-5. Therefore along each of the pair of axes constituting spatial order for the boundary component, there would be an availability of a measuring-rod 1 + 2 + 3 + 4 + 5 + 6.

This availability of the measuring-rod 1 + 2 + 3 + 4 + 5 + 6 along each of the axes is parallel to the potentiality of 4-space to play the role of dimension to create 6-space.

With this the potentiality at the boundary of hypercube-5 with seat of the origin at its center and 6-space in the role of origin is of the order of the potentiality of the measuring-rod available within each of the boundary component of hypercube-5.

With this the Trinity of Gods are of the order of un-manifest, on one side of hyper surface, and with potentiality to manifest on the other side of the hyper surface as a pair of cubes coordinated on a di-monad format, which at first instance takes us to hypercube-4 with sealed origin and then as a next step to hypercube-5 with unlocking of the seal of the origin of hypercube-4.

This pair of steps is parallel to first covering the spatial order and then as a next step covering the solid order. This simultaneous handling of spatial and solid orders is the simultaneous handling of a square and a cube.

The simultaneous handling of SQUARE (81) and CUBE (31) is to simultaneously handle NVF 112 which is precisely the NVF (MATHEMATICS).

Space Book : Chapter Order Four

Here the aspect being focused is as to how the knowledge has been sequenced and preserved. The sequential order of formulations FIRST, SECOND, THIRD and so on are sequencing the chase.

First
72
Origin

First in order is the origin. The artifice 72 is of the Order 6 x 12. This is of the order of 72 coordinates requirement for fixation of boundary of 6-space of 12 components.

$a^6 : 12b^5$

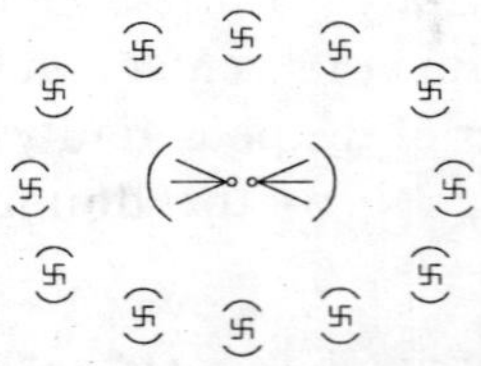

Second = 60 = order = four

Order itself is second in order as order always presupposes the existence of first to which it is to be second. The artifice 60 accepts organization as 5 x 12, which is equivalent to the requirement of twelve five-space component (hypercubes-5) constituting the boundary of 6-space

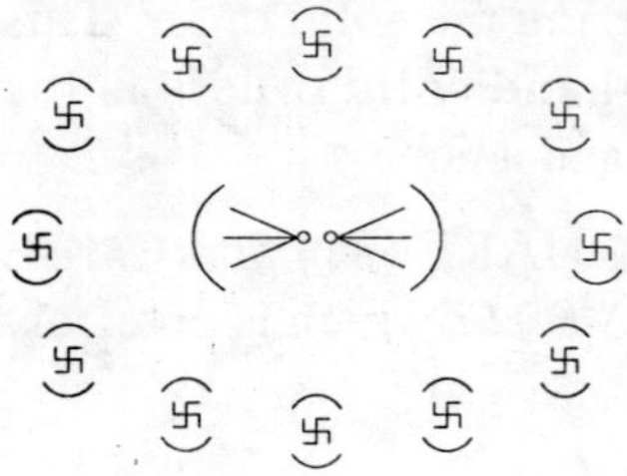

Domain 6-space

4-space dimension

5-space Boundary

Third = 59 = solid

5-space is of solid dimensional order. As such, third in order, is solid, as dimensional order at the boundary. First in order is the domain space supplying coordinates for fixation of boundary. Second in order is the set up at the boundary. And in A sequence, the third in order is the solid dimensional order itself of the space constituting the boundary

First	72	origin
Second	60	four
Third	59	solid

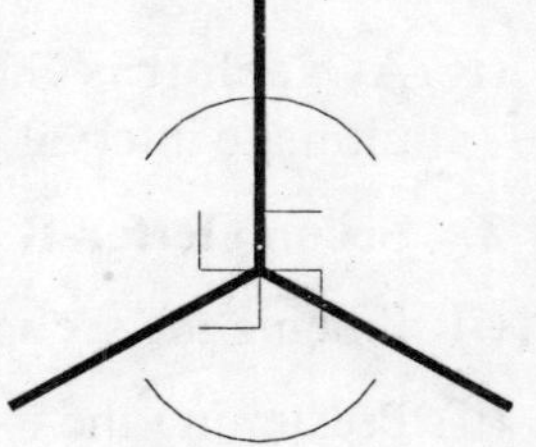

Fourth = 88 = volume

First	72 = 6 x 12	origin	6-space coordinating its boundary
Second	60 = 5 x 12	four	4-space as dimensional order of 6-space (3, 4, 5, 6) manifestation layer with 6- space as origin fold and 4-space as dimension fold
Third	59	solid,	5-space as solid dimensional order space (5-space as domain fold of solid order)

Fourth 88 volume 3-space/volume as dimensional order

Dimension	boundary	domain	origin
Fold	fold	fold	fold

Fifth = 49 = axes

1) First letter-A

(i) It being first, its form and formulations, is of unique features of everything what is to be first.

(ii) It is the first axis of the source, square.

(iii) It is the first feature of the ultimate -interval, as a mind lock and key.

(iv) As such it is the form settled by a pair of spheres with ultimate interval as a mind a lock as first axis.

2. Second letter-B

(i) It being the second, presumes the existence of the first.

(ii) Presuming the existence of the first, it is the second axis of the source square.

(iii) It is the ultimate interval, as a mind lock and key, as second axis.

(iv) As such it is the form settled by a pair of spheres along the second axis.

(v) It being the second axis, and the second letter, so the artifice of whole number 2 also comes into play, for the reason of the square as source, a spatial order of square domain, and further 2 being whole number, also having the feature of wholeness of 2 as 1, therefore the form of the second letter has the feature of 2 as 1, and 1 as of 2 parts, and each part a complete whole and accordingly the pair of spheres, work out the form with a pair of half spheres, one half sphere to be availed for each of the 2 spheres.

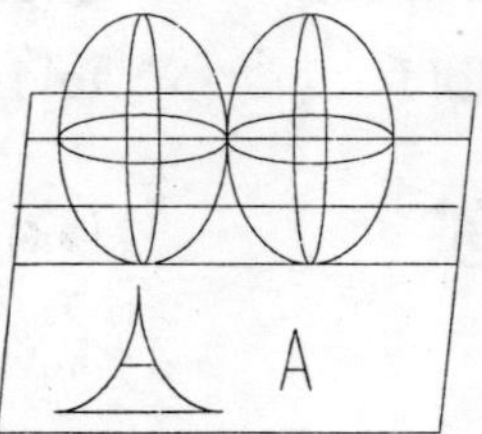

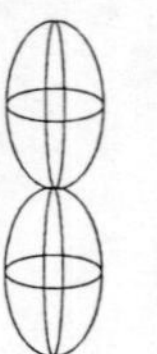

Sixth = 80 = creator

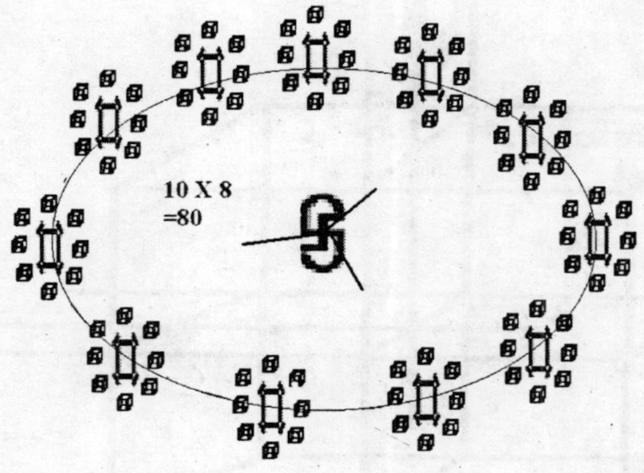

Seventh = 93 = entity

93 = 31 + 31 + 31

Entity = cube + cube + cube

NVF (solid order) = NVF (linear order)

Eighth = 57 = height

Height end = 90 = vertical = surfaces

Horizontal = 118 = 59 + 59 = solid + solid

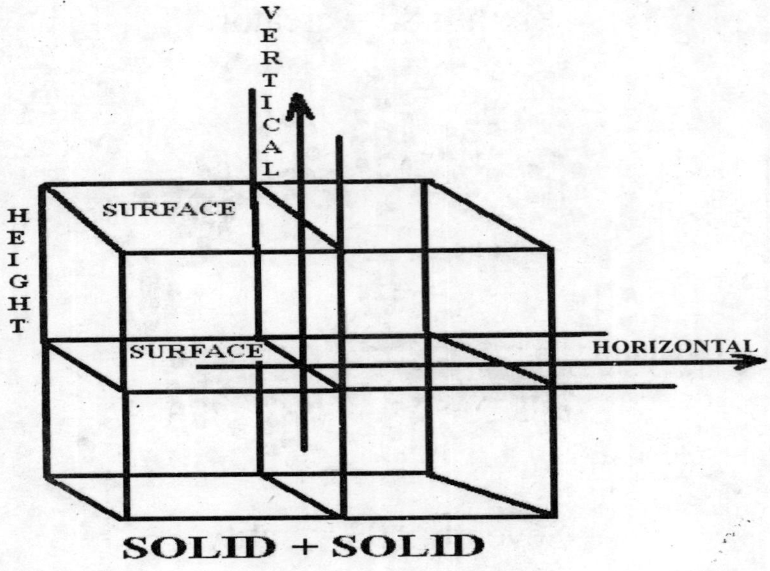

Ninth = 65 = center

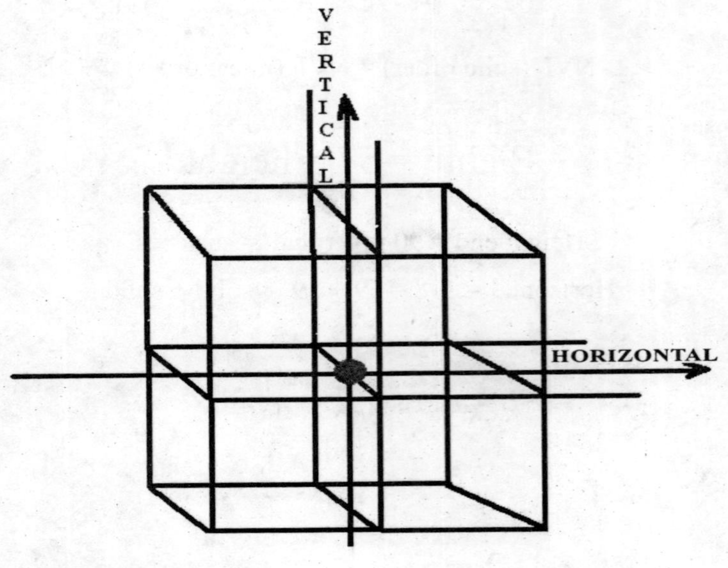

Tenth = 67 = water

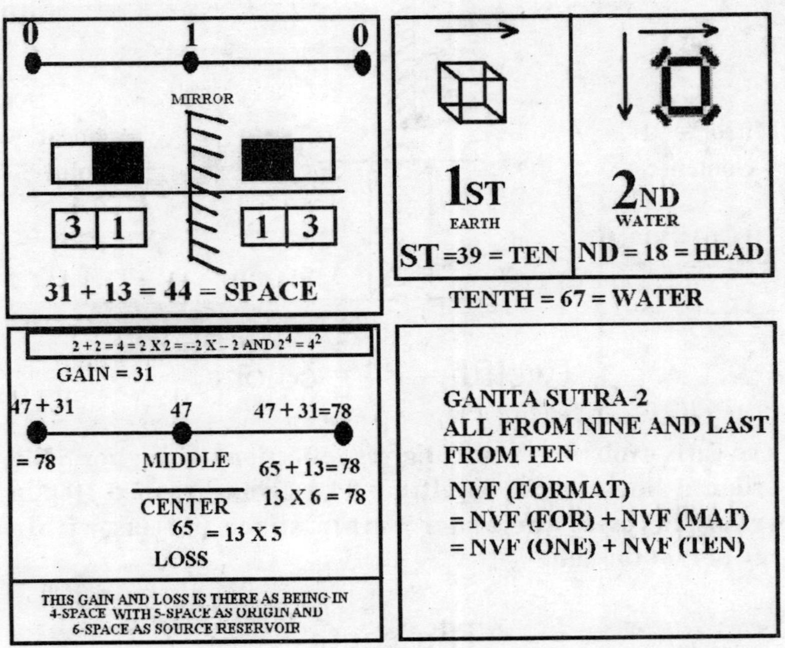

Eleventh = 91 = mirror

The format with artifices pairing of 01 and 10 makes them a reflection pair. This, this way, takes to the joint of the format as of mirror Properties. As such, the sequential order from the tenth stage takes to a step ahead of the order as mirror being the eleventh step in the order.

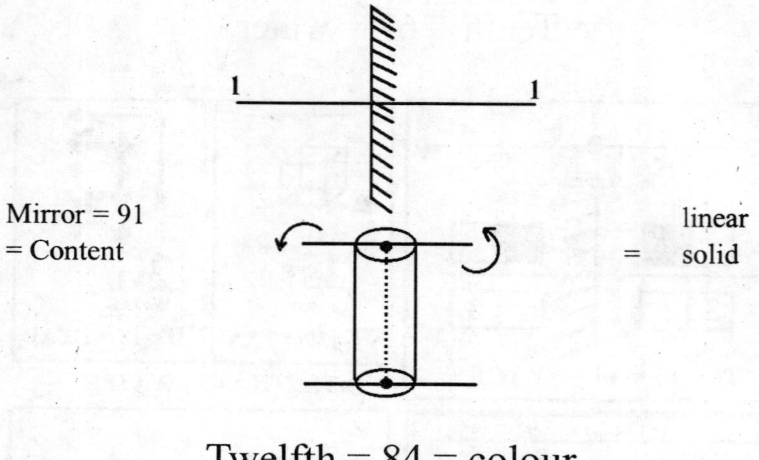

Twelfth = 84 = colour

Eleventh = mirror = content gives reflection pairing for Solid order as linear order Twelfth = 84 = colour, as next step in order, takes to the order within space (44) as within ray (44) of the sun.

Thirteenth
= thir (55) + teenth (44+28)
= Sky + origin

Here there is a shift to sky with an origin, as a first sequential order step of the seven colours spectrum within ray equipped with mirror and colours.

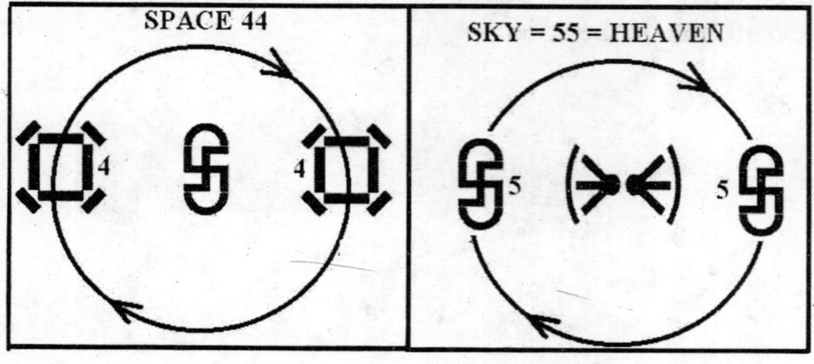

The seven colour beams in the order VIBGOUR takes through the formulations THIRTEENTH to NINETEENTH. The organizational stitches of these formulations are (i) 23 (end) (ii) 2 (60 = TWO + 2) (iii) 19 (which makes beam/edge as beams/ edges) (iv) 3 (Earth-Lord) (v) 5 (center to stop) (vi) 44 (Black to format) and (vii) 10 (cone-angle)

Thirteenth	= 55 + origin	violate (i)
23	= sky + origin	format
Fourteenth	= 60 + origin	indigo (ii)
2	= four + origin	two
Fifteenth	= 21 + origin	blue (iii)
19	= beam/Edge + origin	mind
Sixteenth	= 52 + origin	green (iv)
3	= earth + origin	lord
Seventeenth	= 65 + origin	orange (v)
5	= center + origin	stop
Eighteenth	= 29 + origin	ultra (vi)
44	= black/see + origin	format
Nineteenth	= 37 + origin	red (vii)
10	= cone/seal + origin	code

*****

Section IV

SUN GOD CREATOR

FOUR HEAD LORD LABORATORY

I

SPACE BOOK
44, 43
WEAVE
23 + 5, 1, 22 + 5
WE 0, 23, 5
AVE 1, 22, 5
FIRST BIBLE GENERIC
NVF 6
AE 0, 1, 5

II

SUTRA................THREAD

सूत्रः 56

LIGHT
DOMAIN
FLOW
(AIR, AIR)

III

READ AIR

AIR IS THE FOURTH ELEMENT

FOUR + 4 = 64 = NVF(ZERO)

FOUR – 4 = 56 = LIGHT, FLOW
DOMAIN,
(AIR, AIR)

IV

ONE

34

ONE + ONE = 34 + 34 = 68 = JOINT

ONE + 1 = 35 = EYE

ONE – 1 = 33 = SEED

ONE + 2 = 36 = REAL

ONE – 2 = 32 = LIFE

ONE + 3 = 37 = CONE

ONE – 3 = 31 = CUBE

ONE + 4 = 38 = FIRE

ONE – 4 = 30 = BIBLE

V

TWO

58

TWO + TWO = 116 = ETERNITY

TWO + 1 = 59 = SOLID

TWO – 1 = 57 = SLEEP

TWO + 2 = 60 = FOUR

TWO – 2 = 56 = LIGHT

TWO + 3 = 61 = GENERIC

TWO – 3 = 55 = SKY

TWO + 4 = 62 = COSMIC

TWO – 4 = 54 = SUN

VI
THREE
56

THREE + THREE = 112= MATHEMATICS

THREE + 1 = 57 = SLEEP

THREE – 1= 55 = SKY

THREE + 2 = 58 = TWO

THREE – 2= 54 = SUN

THREE + 3 = 59 = SOLID

THREE – 3= 53 = NOSE

THREE + 4 = 60 = PURE

THREE – 4 = 52 = EARTH

VII

FOUR

60

FOUR + FOUR =12O = FAITH FULLY

FOUR + 1 = 61 = CHURCH

FOUR – 1 = 59 = SOLID

FOUR + 2 = 62 = COSMIC

FOUR – 2 = 58 = TWO

FOUR + 3 = 63 = LIMIT

FOUR – 3 = 57 = SLEEP

FOUR + 4 = 64 = ZERO/UNIT

FOUR – 4 = 56 = LIGHT

VIII

FIVE

42

FIVE + FIVE = 84 = COLOUR

FIVE + 1 = 43 = BOOK

FIVE – 1 = 41 = AFFINE

FIVE + 2 = 44 = SPACE

FIVE – 2 = 40 = MIND

FIVE + 3 = 45 = RANGE

FIVE – 3 = 39 = ANGLE

FIVE + 4 = 46 = LOGIC

FIVE – 4 = 38 = FIRE

IX

SIX

52

SIX + SIX = 104 = SUBSTANCE

SIX + 1 = 53 = NOSE

SIX – 1 = 51 = FULL

SIX + 2 = 54 = SUN

SIX – 2 = 50 = CIRCLE

SIX + 3 = 55 = SKY/HEAVEN

SIX – 3 = 49 = LORD

SIX + 4 = 56 = LIGHT

SIX – 4 = 48 = BLOOD

X

SEVEN = 65 = CENTER

SEVEN + SEVEN = 130 = SUN RISING

SEVEN + 1 = 66 = FAMILY

SEVEN – 1 = 64 = ZERO

SEVEN + 2 = 67 = WATER

SEVEN – 2 = 63 = LIMIT

SEVEN + 3 = 68 = JOINT

SEVEN – 3 = 62 = COSMIC

SEVEN + 4 = 69 = ZOOM

SEVEN – 4 = 61 = CHURCH

XI

EIGHT = 49 = LORD

EIGHT + EIGHT = 98 = TRUST

EIGHT+ 1 = 50 = VOID

EIGHT – 1 = 48 = BLOOD

EIGHT + 2 = 51 = FULL

EIGHT – 2 = 47 = MONAD

EIGHT + 3 = 52 = HEART

EIGHT – 3 = 46 = LOGIC

EIGHT + 4 = 53 = NOSE

EIGHT – 4 = 45 = RANGE

XII

NINE = 42 = FIVE

OLD = 31 = CUBE = 3-SPACE

NEW = 42 = HYPER CUBE = 4-SPACE

FIVE VOWELS AND GEOMETRIES OF 4-SPACE

A = 1 = FIRST GEOMETRY

E = 5 = 5 NON NEGATIVE GEOMETRIES

I = 9 = TOTAL GEOMETRIES

O = 15 = 1 + 2 + 3 + 4 + 5

U = 21 = 1 + 2 + 3 + 4 + 5 + 6

XIII

TEN = 39 = ANGLE

TEN + TEN = 78 = AMBROSIA

TEN+ 1 = 40 = MIND

TEN – 1 = 38 =FIRE

TEN + 2 = 41 = AFFINE

TEN – 2 = 37 = CONE

TEN + 3 = 42 = NEW

TEN – 3 = 36 = REAL

TEN + 4 = 43 = BOOK

TEN – 4 = 35 = EYE

XIV

ELEVEN = 63 = LIMIT

ELEVEN + ELEVEN = 126

= CONTINUING/LIMIT MEANING/BIBLE SEED

ELEVEN + 1 = 64 = ZERO/UNIT

ELEVEN – 1 = 62 = COSMIC

ELEVEN + 2 = 65 = CENTER

ELEVEN – 2 = 61 = CHURCH

ELEVEN + 3 = 66 = FAMILY

ELEVEN – 3 = 60 = PURE

ELEVEN + 4 = 67 = WATER

ELEVEN – 4 = 59 = SOLID

5-SPACE SPACE GEOMETRIES

ELEVEN

XV

TWELVE = 77 = MATTER

TWELVE + TWELVE = 154

= TRANSFORMING

= WORSHIPPING

TWELVE + 1 = 78 = AMBROSIA

TWELVE–1=76 = HIGHEST

TWELVE + 2 = 79 = NATURE

TWELVE – 2 = 75 = ATERNAL

TWELVE + 3 = 80 = CREATOR

TWELVE – 3 = 74 = PAIRING

TWELVE + 4 = 81 = SQUARE

TWELVE – 4 = 73 = FORMAT

CHRIST ACCEPTS SPACE (44) SEED (33) AS GENERIC NVF 77

XVI

THIRTEEN = 99 = THOUGHT

THIRTEEN + THIRTEEN = 198

= LOVING KINDNESSES

THIRTEEN + 1 = 100 = EXCELLENT

THIRTEEN – 1 = 98 = CRYSTAL/EYE SIGHT

THIRTEEN + 2 = 101 = INTERVAL/ULTIMATE

THIRTEEN – 2 = 97 = FOUR FOLD/SUPREME

THIRTEEN + 3 = 102 = SEVEN FOLD/PRAYERS

THIRTEEN – 3 = 96 = SIXTEEN

THIRTEEN + 4 = 103 = COUNTING

THIRTEEN – 4 = 95 = DISCOVER/PROCESS/
PROMISE/PURIFY/RENEWING

THOUGHT (99) IS TO BE ABOUT
SPACE (44) HEAVEN (55)

XVII

FOURTEEN = 104 = FOUR SPACE

FOURTEEN + FOURTEEN = 208
= (EARTH, EARTH, EARTH, EARTH)

FOURTEEN + 1 = 105 = 5 x (1 + 2 + 3 + 4 + 5 + 6)

FOURTEEN – 1 = 103 = COUNTING

FOURTEEN + 2 = 106 = HONESTY/TRUTHS

FOURTEEN – 2 = 102 = SEVEN FOLD/PRAYERS

FOURTEEN + 3 = 107 = VISIONS/TRUSTED/
UTTERANCE/STRIPPED

FOURTEEN – 3 = 101 = INTERVAL/ULTIMATE

FOURTEEN + 4 = 108 = ARTIFICERS/
FULFILLING/TWILIGHT

FOURTEEN – 4 = 100 = EXCELLENT

XVIII

FIFTEEN = 65 = SEVEN

1. (65, 56) REFLECTION PAIR

2. 65 = 5 x 13, A TAGGING AND

   COORDINATION OF 13 EDGED HYPERCUBE–4 WITH CENTER OF 4-SPACE AS A SEAT OF 5-SPACE IN THE ROLE OF ORIGIN FOLD.

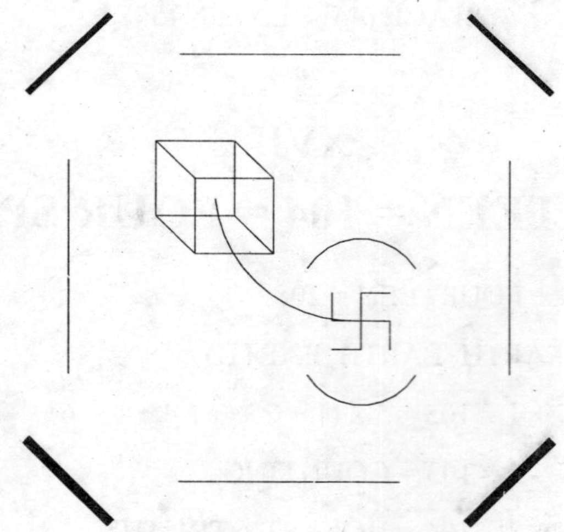

XIX

SIXTEEN = 96 = EARTH SPACE

96 = 6 x 16

6 x SIXTEEN = 6 x 96 = 6 x 6 x 16

6 x 6 x 16 = 6 x 6 x SIXTEEN = 6 x 6 x 6 x 16

AND THE PROCESS CONTINUES AS A

SEQUENTIAL FLOW FROM CENTER OF PENTAGON

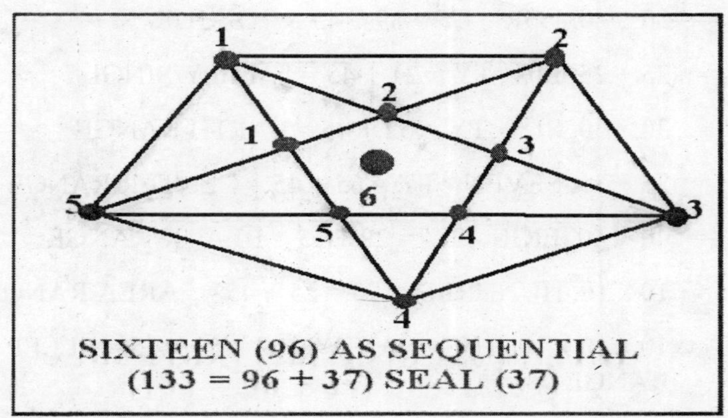

XX

SEVENTEEN = 109 = BLACK CREATOR BLACK SUN GOD

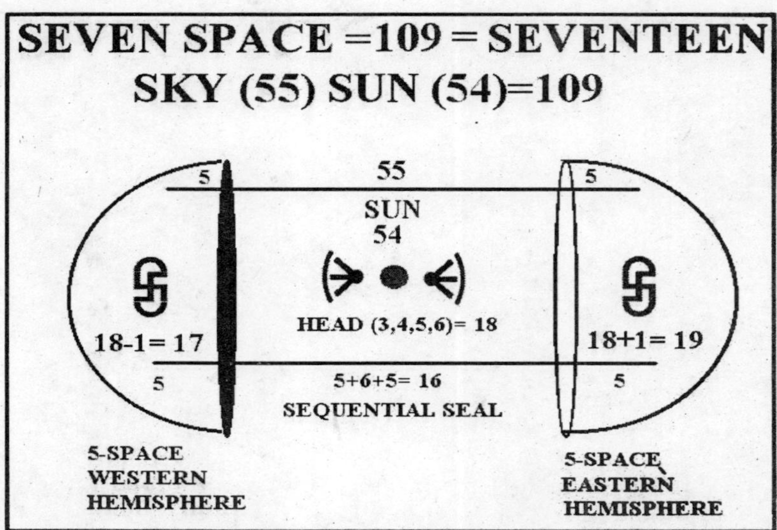

20 = 10 + 10, TWEN - TY = 52 + 45 = EARTH RANGE

30 = 15 + 15, THIR - TY = 45 + 45 = (RANGE, RANGE)

40 = 20 + 20, FOR - TY = 39 + 45 = COLOUR

50 = 25 + 25, FIF - TY = 21 + 45 = FAMILY/SINGLE

60 = 30 + 30, SIX - TY = 52 + 45 = EARTH RANGE

70 = 35 + 35, SEVEN - TY = 65 + 45 = CENTER RANGE

80 = 40 + 40, EIGH - TY = 29 + 45 = BLACK RANGE

100 = 10 x 10, HUNDRED = 70 = 25 + 45 = AREA RANGE

1000 = 10 x 10 x 10, THOUSAND = 102 = 57 + 45 = SLEEP RANGE

*****

Section V

SHRIMAD BHAGWAD GEETA

SHRIMAD BHAGWAD GEETA

Divine song, *Shrimad Bhagwad Geeta*, accepts cosmic organization format which manifests with manifestation of *Divya Jyoti* flowing through rays of the Sun from orb of the Sun into the domain of Earth. It deserves to be approached on tetra monad format as creator's Space (4-Space) playing the role of dimension of *Vishnu Lok* (6-Space). It is going to be a multi-volume exercise. However here for an initial glimpse of it is being drawn the first outline as tabulated information as under:

Geeta Study Zone Chase Step-1

Srimad Bhagwad Gita Study – Zone										
$\frac{a}{2}$	$\frac{a^2}{4a}$	$\frac{a^3}{6a^2}$	$\frac{a^4}{8a^3}$	$\frac{a^5}{10a^4}$	$\frac{a^6}{12a^5}$	$\frac{a^5}{10a^4}$	$\frac{a^2}{8a}$	$\frac{a^3}{6a^2}$	$\frac{a^4}{4a^3}$	___
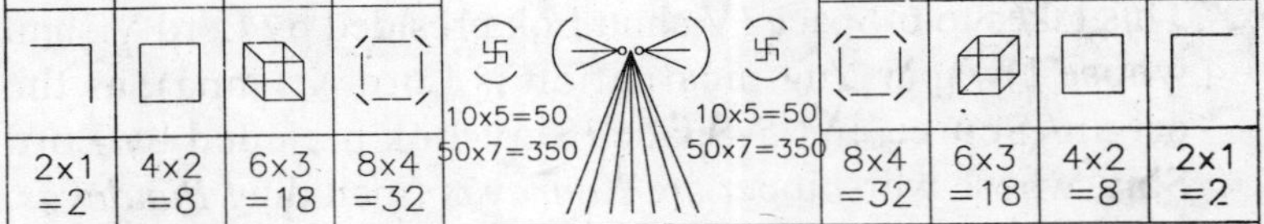				10x5=50 50x7=350		10x5=50 50x7=350				
2x1 =2	4x2 =8	6x3 =18	8x4 =32				8x4 =32	6x3 =18	4x2 =8	2x1 =2
Orbitals				350+350=700						
2	6	10	14	18=5+6+7						

1. Geeta study zone is the inner most fold of the transcendental worlds. This is manifesting as a point reservoir (Bindu Sarovar) at the joints of di-monad.

2. This manifestation at dimension of the dimension level of the dimensional order of di-monad format may permit expression as an interval of two units with joint as the Bindu Sarovar (point reservoir state of the study zone).

3. This state in its full manifestation expression shall be taking us to a state where each of two parts of di-monad (as unit intervals, as representative regular bodies of 1-Space while being in the role of dimension of dimension), shall be manifesting first as 3-Space and then as 5-Space. In the

process the zero state joint (Bindu Sarovar/point reservoir) getting full expression for its complete manifestation from the state of zero space in the role of dimension of dimension, firstly as 2-Space to 4-Space and then from 4-Space to 6-Space as the reservoir at the middle of the manifestation.

4. Geometrically this state of manifestation of di-monad format may be chased as a pair of hypercubes-5, as representative regular bodies of 5-Space, as pair of manifested parts of di-monad and hyper cube-6 as representative regular body of 6-Space as the joint of the di-monad.

5. Here in this state the above manifestation of di-monad format emerges to be a format of space (fifth Mahabhut-Akash) with Sun (Surya/Pursha/Atman as Sixth element transcending Panch Mahabhut) as the source reservoir for all transcendental flow for manifested existence as Brahmand (creations domain of Lord Brahma, Creator the Supreme).

6. This takes to 6-Space (Vishnu Lok presided by Lord Vishnu whose Dwapar Yug incarnation is Lord Krishna), as the origin/ Source and 5-Space (Shiv-Lok presided by Lord Shiv, whose worshipper are *Kauravas* (including *Pandavas*, and amongst them *Arjuna* being the *Sakha* of Krishna) as the space within which the existence phenomena is subject matter of chase of divine dialogue (of Lord Krishna and Arjuna): Shrimad Bhagwad Geeta, and as such the Space (Akash tatav) with Sun (Surya as Atman/self) emerges to be the Study Zone of Shrimad Bhagwad Geeta.

*****

Geeta chapter - 1
अर्जुनविषादयोग

ArjunVishad Yoga
Chasing Organization on Geometric Format

ARJUNA VISHAD YOG

SR	ASPECT	TITLE	UVACHA	TEXT	PUSHIPIKA	TOTAL
1.	PADAS	3	12	558	13	586
2.	AKSHRAS	7	37	1504	48	1596

प्रथमोऽध्यायः अर्जुनविषादयोग श्लोकानिः ४७

Prathmo Adhyay: ArjunVishad Yoga: Shalokas 47

पुष्पिकाः– (इति श्री महाभारते शतसाहस्र्यां वैयासिक्यां भीष्मपर्वणि) (ॐ तत्सदिति) श्रीमद्भगवदगीता सूपनिषत्सु ब्रह्मविद्यायां योगशास्त्रे श्रीकृष्णार्जुनसंवादे **अर्जुनविषाद** योगो नाम **प्रथमो**ऽध्यायः ।।१।।

Pushpika: - (iti shri Mahabharte shatsahsatram vaiyasikyam bhishampariyani) (Om tatsadity) (Shrimad Bhagwad Geeta Suupnishatsu Brahamvidyayam yogshastre Shri Krishan Arjun Sambade **Arjun Vishad** Yogo nam **Prathmo** Adhyay)

Outline

Section-0 Continuity Background

1.1.1 Continuity background

1.1.2 Transcendental worlds

1.1.3 Lord Shiv

1.1.4 Lord Shiv's Family

1.1.5 Inner most fold of transcendental worlds

1.1.6 Transcendence from transcendental world to hyper transcendental worlds.

1.1.7 Kauravas: Worshipper of Lord Shiv.

1.1.8 Lord Shiv is Lord of Lord Vishnu and Lord Vishnu is Lord of Lord Shiv.

Section-1. General

1.1.1 Shaddarshan–six ways to glimpse the origin.

1.1.2 Asth-Dash Purans–eighteen ways to retain glimpsed origin.

1.1.3 Asth-Dash ways to transcend Vishwa to be in creator's space.

1.1.4 Geeta as Scripture of Asth-Dash ways for attaining unison with Braham, the ultimate.

1.1.5 Vishad as is the first aspect of Yoga, the discipline for attaining unison with the ultimate.

1.1.6 Vishad, Udasi and Vairagya.

Section-2. Urge to know and chase organization of Geeta Chapter-1.

1.1.1 Within 6-Space on geometric format accepting measuring rod constituted by representative regulars bodies of 1 to 6 Space.

1.1.2 As orbitals within orbits parallel to electronic configuration of atoms of elements.

1.1.3 As monad with placement at the middle.

1.1.4 As initiation from the seat of origin as seat of compactification of Spaces.

1.1.5 Dissolution of dimensional order of the space with unlocking of the seal of the origin of dimensional frame of the space.

1.1.6 From manifest (3-Space) to unmanifest domain (4-Space).

1.1.7 Life and existence at micro level (within 4-Space).

1.1.8 Re-manifestation of unmanifest domain.

1.1.9 Transcendence from unmanifest (*Avyakto*), existence order of 4-Space, to base of unmanifest (*Avyakto-Avyaktat*), existence order of 5-Space reality.

1.1.10 Pursha as phenomena of hyper transcendental existence order of 6-Space Reality.

Section-3. Urge to know and chase "Arjun Vishad Yog"

1.1.1 Arjuna's Vishad.

1.1.2 Dissolution of body and continuity of life.

1.1.3 Manifestation of body only at the middle.

1.1.4 Lord as source of emanation of life.

1.1.5 Life process can be approached within body only at the middle of the process.

1.1.6 Sarav-Arambh parityagi.

1.1.7 One remains undefined and "one more than previous one" is to be the first rule.

1.1.8 From "one more than previous one" to "one less than the previous one" there is a fourteen steps long gap-range in between.

General

One of the scripture's command is: "*Saravarambh Parityagi*"; no beginning, no beginning point of the processing: every processing here is to be from the middle, as all processes have their beginning already having emanated from the lord as source. The reality of over existence is that, one is in the existence queue somewhere in between, as his father and forefather having existed prior to him and his son and grandson and other in the lineage worked follow. One exists **(born)** as such in between, so one is here at the middle. It is from middle to the end becomes the order of chase. To start from the middle become the rule.

To start from the middle

It is on the tenth day of epic war that Bhishma had fallen from his chariot in the battlefield of Kurukshetra. It is at this stage that Sanjaya the charioteer of Dhritrashtara had returned from the battle field and joined Dhritrastra at Hastinapur and it is at this occasion, in the middle of the war that Dhritrastra had expressed the desire to hear all from the beginning of the war everything with details from Sanjaya.

Middle to be out, at the end or at the beginning

For location, as a process of organization, the middle which otherwise is in between the beginning and end, is to be out, as after the end or as before the beginning. It is there being so, as the process of fixation, if fixes the beginning point as first step, it is to fix the end point as a second step and it is only at the third step that the middle would stand fixed.

This sequence of fixation as first step (as the fixation of beginning), as second step (the fixation of end of the range) and only as third step, there being a fixation of middle point, in that sequence and order would make processing steps placements as first, second and third and middle being the third step. So it is after the end point.

The exhaustive coverage processing of the creator's space (4-Space) where reversal of orientation (because of $2 + 2 = 2 \times 2 = -2 \times -2$) is permissible would make processing steps as first step as end point, second step as the beginning of the range and then third step as fixation of the middle. With it the arrangement of placements would be 3, 2, 1 and the middle as third step would be even prior to the beginning point of the range.

It is this unique feature of processing process of the creator's space which with its characteristics' feature as to permit reversal of orientation and even to reach prior to the beginning of the range, which is the transcendental feature which makes the reality of creator's space in the role of dimension of *Vishnu Lok*.

This transcendental feature of the creator's space while in the role of dimension of *Vishnu Lok* makes the organization potentialities even at the dimensional level of *Vishnu Lok* as of the order which sustains even reversal of orientations within the dimensional order.

It is this transcendental feature of sustaining reversal of orientation within the dimensional order because of which the dimension itself as domain makes the process lively at its dimensional level that is at the level of dimension of dimension.

It is because of this transcendental feature of making the processing lively at the level of dimension of dimension and that to with reversal of orientation that the linear order gets transcendently transform as manifestation order as much as that 1-Space transcendently transform as 1-manifestation layer with 1-Space as domain fold, and –1 Space as the dimension fold thereof.

The other, two folds, 0-Space as boundary fold and 2-Space as origin fold also being inter-manifested as 0-Space as dimension of 2-Space, makes the organization of manifestation fold as a pair of organization inter-manifesting (-1 Space, +1 Space) and (0-Space, 2-Space) and thereby the whole process of linear order getting transcendently transform as self-referral spatial order and there emerging orbitals within in the orbits with middle of the range as the compactification point of nuclei and it is the transcendence through it that the ultimate is to be attained.

It is this which settles the ultimate knowledge zone as study zone of Shrimad Bhagwad Geeta and it is this for whose enlightenment is the question of Dhritrashtra to Sanjaya as that

VERSE-1

धृतराष्ट्र उवाच *dhritrastra uvacha*

धर्मक्षेत्रे कुरूक्षेत्रे समवेता युयुत्सवः ।

dharma-ksetre kuri-ksetre samaveta yuyutsavah

मामकाः पाण्डवायश्चैव किमकुर्वत सञ्जय ।। १ ।।

mamakah pandavas caiva kim akurvata sanjaya

धर्मक्षेत्रे	कुरूक्षेत्रे	समवेता	युयुत्सवः ।
dharma-ksetre	*kuri-ksetre*	*samaveta*	*yuyutsavah*

मामकाः	पाण्डवाः	च	एव	किम	अकुर्वत	संजय
Mamakah	*pandava*	*cha*	*aiva*	*kim*	*akurvata*	*sanjaya*

Dhritrashtra said:

Assembled on the field of Dharma, O Sanjaya, on the field of the Kurus, eager to fight, what did my people and the Pandavas do?

Initiation for transcendence at the middle

Dhritrashtra, King of Kauravas clan of worships us of Lord Shiv, was blind and he despite eyes closed was inherently capable of transcending manifestations to be at their base and to fathom through the re-cycling processes of transcendental base and as such could reach uptil the happening of the tenth day of the war but was naturally held up there and wanted re-evaluation of everything again to transcend further from this tenth day state as the middle state and to project and fathom of the following range of transcendental pilgrimage as it was distinct as end of the war and existence beyond there to.

It is this initiation for transcendence at the middle for which Dhritrashtra has the urge and it is this transcendental pilgrimage through which this scripture enlightens.

This initiation for transcendence at the middle as such is the initiation of the chase of the urge to be translated as a transcendence pilgrimage through the middle as the origin which with its transcendental seal is segregating the initial ten days war range from the following eight days war range, which otherwise are compactified at the same point unfolding simultaneously like ten *Mandals* and eight *Astaks* unfolding of *Rig Ved* range.

As such the initiation question of Verse-1 is the question for re-knowing and re-evaluation as well as for initial unfoldment of ten *Mandals* organization, as well as this initiation question of Verse-1 as put after tenth day of war, is the question as to be re-put for knowing and enlightenment about the unfoldment process of eight *Ashtaks* organization of *Rig Ved*.

This is the urge of the initiation question which better could be appreciated only when it was put at the middle of the war range like the question of simultaneous emergence of ten *Mandalas* organization and eight *Astaks* organization.

This is the urge of the initiation question which as an organization of the transcendental worlds of solid dimensional order having full expression along all the ten directions parallel to ten *Mandals* expression fully enveloping the solid dimensional order with its further potentialities to unfold its inner most fold as of hyper solid dimensional order for the *Ati-Vahkas* to carry through the transcendental world to the hyper transcendental worlds of *Vishnu Lok* as 6-Space reality having potentialities in its dimensional order to be of the order of *Asth-Prakriti* (8-Space reality).

Text: Organization within orbits as orbitals

Orbit-1

Orbital-1 Verses 1 and 2

1. Verse-1 Basic question raised by Dhritrastra
2. Verse-2 Sanjaya having taken above question for answer

Orbit-2

Orbital-1 Verse 3 and 4

1. Verse-3 Duryodhan goes to Dronacharya
2. Verse-4 Duryodhan starts describing and gives first names of Maharathis of Pandava's army.

Orbital-2 Verses 5, 6, 7, 8, 9 and 10

1-6 Verses 5-10 Duryodhna completes the descriptions of Maharathis of both army and concludes the descriptions stating that as that while army of his side was fully potentialized under the command of Bhishma, and on the other hand the army of the Pandavas was insecure under the command of Bhim.

Orbit-3

Orbital-1 Verses 11 and 12

1. Verse-11 Duryodhna gives command to commanders of his army that they to protect their commander-in-chief Bhishma.

2. Verse-12 Hearing this command, the commander-in-chief Bhishma blows his conch to announce the beginning of the war.

Orbital-2 Verses 13, 14, 15, 16, 17 and 18

1-6 Verses-13-18 There follows blowing of conches from Commanders and everybody from all sides.

Orbital-3 Verses 19, 20, 21, 22, 23, 24, 25, 26, 27 and 28

1-10 Verses-19-28 The sound of conches resonated whole of the battle-field and space and everybody's heart on Kaurva side started echoing with fear and on the other side, on Arjuna had dawn the reality of the war that he was to fight and kill his relatives and he approached Krishna.

Orbit-4

Orbital-1 Verses 29,30

1-2. Verses-29.30 Arjuna expressed before the lord about his state of mind and body as that he was even not in a position to stand.

Orbital-2 Verses 31-36

1-6 Verses-31-36 Arjuna logics out with the lord that he was noticing adverse symptoms of war and he was seeing only sin in killing the relatives.

Orbital-3 **Verses 37-46**

1-10 Verse-37-46 Arjuna further logics out and narrates about the worst consequences of the war and gives his conclusion as that instead of going to war even if he is killed by his opponent even while he being armless would be preferable for him.

Orbital-4 **Verse 47**

1 Verse-47 So, saying, Arjuna throws his bow and arrows and fulfilled with sorrow and desperation sits down in the chariot.

This as such, as first step placement in fourth orbital of fourth orbit, is the first chapters coverage. The chariot of Arjuna was taken to the middle separating line of placements of the armies of the both sides at a point of time of description of the text of chapter-1 precisely at middle of this range which takes us immediately after twenty third verse and before completion of 24th verse as would get fixed the middle of the range of 47 verses range of Chapter-1 in between verses 23 and 24.

In this light, the features may be consolidated and tabulated as under:

Chapter-1		**Text**		**Electronic configuration of silver (Ag 47)**					
1	2	3	4	5	6	7	8	9	10
S1	S2	P2	S3	P3	D3	S4	P4	D4	F4
2	2	6	2	6	10	2	6	10	1
Chapter-1		**Geeta Parinam**			**Electronic configuration of silver (Ag 47)**				
1	2	3	4	5	6	7	8	9	10
S1	S2	P2	S3	P3	D3	S4	P4	D4	F4
2	2	6	2	6	10	2	6	10	1

SH	PADAS	AKSHRAS	CHANDAS	Orbits	Orbitals	UVACHA
1	11=11	32	Anustap	Orbit-1	Orbital-1	Dhritrashtra
2	11=22	32	Anustap	Orbit-1	Orbital-1	Sanjaya
3	11=33	32	Anustap;	Orbit-2	Orbital-1	Sanjaya
4	11=44	32	Anustap	Orbit-2	Orbital-1	Sanjaya
5	11=55	32	Anustap	Orbit-2	Orbital-2	Sanjaya
6	12=67	32	Anustap	Orbit-2	Orbital-2	Sanjaya
7	14=91	32	Anustap	Orbit-2	Orbital-2	Sanjaya
8	15=106	32	Anustap	Orbit-2	Orbital-2	Sanjaya
9	09=115	32	Anustap	Orbit-2	Orbital-2	Sanjaya
10	11=126	32	Anustap	Orbit-2	Orbital-2	Sanjaya
11	12=138	32	Anustap	Orbit-3	Orbital-1	Sanjaya
12	11=149	32	Anustap	Orbit-3	Orbital-1	Sanjaya
13	13=162	32	Anustap	Orbit-3	Orbital-2	Sanjaya
14	14=176	32	Anustap	Orbit-3	Orbital-2	Sanjaya
15	09=185	32	Anustap	Orbit-3	Orbital-2	Sanjaya
16	08=193	32	Anustap	Orbit-3	Orbital-2	Sanjaya
17	12=205	32	Anustap	Orbit-3	Orbital-2	Sanjaya
18	12=217	32	Anustap	Orbit-3	Orbital-2	Sanjaya
19	12=229	32	Anustap	Orbit-3	Orbital-3	Sanjaya
20	10=239	32	Anustap	Orbit-3	Orbital-3	Sanjaya
21	13=252	32	Anustap	Orbit-3	Orbital-3	Arjun
22	12=267	32	Anustap	Orbit-3	Orbital-3	Arjun
23	11=278	32	Anustap	Orbit-3	Orbital-3	Arjun
24	10=288	32	Anustap	Orbit-3	Orbital-3	Sanjaya
25	11=299	32	Anustap	Orbit-3	Orbital-3	Sanjaya
26	14=313	32	Anustap	Orbit-3	Orbital-3	Sanjaya
27	14=327	32	Anustap	Orbit-3	Orbital-3	Sanjaya
28	12=339	32	Anustap	Orbit-3	Orbital-3	Arjun
29	13=352	32	Anustap	Orbit-4	Orbital-1	Arjun
30	16=368	32	Anustap	Orbit-4	Orbital-1	Arjun
31	12=380	32	Anustap	Orbit-4	Orbital-2	Arjun
32	17=397	32	Anustap	Orbit-4	Orbital-2	Arjun
33	16=413	32	Anustap	Orbit-4	Orbital-2	Arjun
34	13=426	32	Anustap	Orbit-4	Orbital-2	Arjun
35	13=439	32	Anustap	Orbit-4	Orbital-2	Arjun

36	14=453	32	Anustap	Orbit-4	Orbital-2	Arjun
37	14=467	32	Anustap	Orbit-4	Orbital-3	Arjun
38	10=477	32	Anustap	Orbit-4	Orbital-3	Arjun
39	11=488	32	Anustap	Orbit-4	Orbital-3	Arjun
40	11=499	32	Anustap	Orbit-4	Orbital-3	Arjun
41	09=508	32	Anustap	Orbit-4	Orbital-3	Arjun
42	11=519	32	Anustap	Orbit-4	Orbital-3	Arjun
43	09=528	32	Anustap	Orbit-4	Orbital-3	Arjun
44	09=537	32	Anustap	Orbit-4	Orbital-3	Arjun
45	11=548	32	Anustap	Orbit-4	Orbital-3	Arjun
46	12=560	32	Anustap	Orbit-4	Orbital-3	Arjun
47	10=570	32	Anustap	Orbit-4	Orbital-4	Sanjaya

*****

SHRIMAD BHAGWAD GEETA

I

TABLE NO. 1

GEETA PARIMAN SHALOKA

Chapter No.	Shalokas of text	Shalokas as per Gita Pariman
1.	47	47
2.	72	76
3.	43	47
4.	42	45
5.	29	31
6.	47	52
7.	30	31
8.	28	30
9.	34	35
10.	42	45
11.	55	63
12.	20	22
13.	34	35
14.	27	30
15.	20	21
16.	24	25
17.	28	30
18.	78	80
18	700	745

SHRIMAD BHAGWAD GEETA

II

TABLE NO. 1A

GEETA PARIMAN SHALOKA

Chapter No.	Shalokas /atomic no.	Shalokas/atomic No.; Gita Pariman
1.	47 (Ag)	47 (Ag)
2.	72 (Hf)	76 (Os)
3.	43 (Te)	47 (Ag)
4.	42 (Mo)	45 (Rh)
5.	29 (Cu)	31 (Ga)
6.	47 (Ag)	52 (Te)
7.	30 (Zn)	31 (Ga)
8.	28 (Ni)	30 (Zn)
9.	34 (Se)	35 (Br)
10.	42 (Mo)	45 (Rh)
11.	55 (Cs)	63 (Eu)
12.	20 (Ca)	22 (Ti)
13.	34 (Se)	35 (Br)
14.	27 (Co)	30 (Zn)
15.	20 (Ca)	21 (Sc)
16.	24 (Cr)	25 (Mn)
17.	28 (Ni)	30 (Zn)
18.	78 (Pt)	80 (Hg)
18	700	745

SHRIMAD BHAGWAD GEETA

III

TABLE NO. 1B

GEETA PARIMAN SHALOKA

Chapter No.	Shalokas /atomic no.	Shalokas/atomic No.; Gita Pariman
1.	47 (Ag)	47 (Ag)
2.	72 (Hf)	76 (Os)
3.	43 (Te)	As in Chapter-1
4.	42 (Mo)	45 (Rh)
5.	29 (Cu)	31 (Ga)
6.	As in chapter-1	52 (Te)
7.	30 (Zn)	As in Chapter-5
8.	28 (Ni)	30 (Zn)
9.	34 (Se)	35 (Br)
10.	As in Chapter-4	As in Chapter-4
11.	55 (Cs)	63 (Eu)
12.	20 (Ca)	22 (Ti)
13.	As in Chapter-9	As in Chapter-9
14.	27 (Co)	As in Chapter-8
15.	As in Chapter-12	21 (Sc)
16.	24 (Cr)	25 (Mn)
17.	As in Chapter-8	As in Chapter-8
18.	78 (Pt)	80 (Hg)
18	700	745

SHRIMAD BHAGWAD GEETA

IV

TABLE NO. 1C

GEETA PARIMAN SHALOKA

Chapter No.	Shalokas /atomic no.	Shalokas/atomic No.; Gita Pariman
1.	20	Ca
2.	21	Sc
3.	22	Ti
4.	24	Cr
5.	25	Mn
6.	27	Co
7.	28	Ni
8.	29	Cu
9.	30	Zn
10.	31	Ga
11.	34	Se
12.	35	Br
13.	42	Mo
14.	43	Tc
15.	45	Rh
16.	47	Ag
17.	52	Te
18.	55	Cs
19	63	Bu
20	72	Hf
21	76	Os
22	78	Pt
23	80	Hg

SHRIMAD BHAGWAD GEETA

V

TABLE NO. 1D

GEETA PARIMAN SHALOKA

Sr. No.	Shalokas /Elements	Sr. No.	Text Frequency	Sr. No.	Gita Pariman Frequency
1	20 (Ca)	1	2	-	-
2	28 (Ni)	2	2	-	-
3	30 (Zn)	*(6)	1	1	3
4	31 (Ga)	-	-	2	2
5	34 (Sc)	3	2	-	-
6	35 (Br)	-	-	3	2
7	42 (Mo)	4	2	-	-
8	45 (Rh)	-	-	4	2
9	47 (Ag)	5	2	5	2
10	21 (Sc)	-	-	6	1
11	22 (Ti)	-	-	7	1
12	24 (Cr)	7	1	-	-
13	25 (Mn)	-	-	8	1
14	27 (Co)	-	-	9	1
15	29 (Cu)	8	1	-	-
16	43 (Tc)	9	1	-	-
17	52 (Te)	-	-	10	1
18	55 (Cs)	10	1	-	-
19	63 (Eu)	-	-	11	1
20	72 (Hf)	11	1	-	-
21	76 (Og)	-	-	12	1
22	78 (Pt)	12	1	-	-
23	80 (Hg)	-	-	13	1

SHRIMAD BHAGWAD GEETA

VI
TABLE NO. 1E

GEETA PARIMAN SHALOKA

Sr. No.	Shalokas /Elements	1 s	2 s	2 p	3 s	3 p	3 d	4 s	4 p	4 d	4 f	5 s	5 p	5 d	5 s
1	20 (Ca)	2	2	6	2	6	2								
2	21 (Sc)	2	2	6	2	6	1	2							
3	22 (Ti)	2	2	6	2	6	2	2							
4	24 (Cr)	2	2	6	2	6	5	1							
5	25 (Mn)	2	2	6	2	6	5	2							
6	27 (Co)	2	2	6	2	6	7	2							
7	28 (Ni)	2	2	6	2	6	8	2							
8	29 (Cu)	2	2	6	2	6	10	1							
9	30 (Zn)	2	2	6	2	6	10	2							
10	31 (Ga)	2	2	6	2	6	10	2	1						
11	34 (Se)	2	2	6	2	6	10	2	4						
12	35 (Br)	2	2	6	2	6	10	2	5						
13	42 (Mo)	2	2	6	2	6	10	2	6	5	1				
14	43 (Te)	2	2	6	2	6	10	2	6	5	2				
15	45 (Rh)	2	2	6	2	6	10	2	6	8	1				
16	47 (Ag)	2	2	6	2	6	10	2	6	10	1				
17	52 (Te)	2	2	6	2	6	10	2	6	10	2	4			
18	55 (Cs)	2	2	6	2	6	10	2	6	10	2	6	1		
19	63 (Eu)	2	2	6	2	6	10	2	6	10	7	2	6	2	
20	72 (Hf)	2	2	6	2	6	10	2	6	10	14	2	6	2	2
21	76 (Os)	2	2	6	2	6	10	2	6	10	14	2	6	6	2
22	78 (Pt)	2	2	6	2	6	10	2	6	10	14	2	6	9	1
23	80 (Hg)	2	2	6	2	6	10	2	6	10	14	2	6	10	2

SHRIMAD BHAGWAD GEETA

VII

TABLE NO. 1F

GEETA PARIMAN SHALOKA

Sr. No.	Shalokas /Elements	Electronic Configuration					
		1	2	3	4	5	6
		S	S+p	S+p+d	S+p+d+f	S+p+d+f	S+...
1	20 (Ca)	2	2+6	2+6+0	2+0+0+0	0+0+0+0	0
2	21 (Sc)	2	2+6	2+6+1	2+0+0+0	0+0+0+0	0
3	22 (Ti)	2	2+6	2+6+2	2+0+0+0	0+0+0+0	0
4	24 (Cr)	2	2+6	2+6+5	1+0+0+0	0+0+0+0	0
5	25 (Mn)	2	2+6	2+6+5	2+0+0+0	0+0+0+0	0
6	27 (Co)	2	2+6	2+6+6	2+0+0+0	0+0+0+0	0
7	28 (Ni)	2	2+6	2+6+8	2+0+0+0	0+0+0+0	0
8	29 (Cu)	2	2+6	2+6+10	1+0+0+0	0+0+0+0	0
9	30 (Zn)	2	2+6	2+6+10	2+0+0+0	0+0+0+0	0
10	31 (Ga)	2	2+6	2+6+10	2+1+0+0	0+0+0+0	0
11	34 (Se)	2	2+6	2+6+10	2+4+0+0	0+0+0+0	0
12	35 (Br)	2	2+6	2+6+10	2+5+0+0	0+0+0+0	0
13	42 (Mo)	2	2+6	2+6+10	2+6+5+0	1+0+0+0	0
14	43 (Te)	2	2+6	2+6+10	2+6+5+0	2+0+0+0	0
15	45 (Rh)	2	2+6	2+6+10	2+6+8+0	1+0+0+0	0
16	47 (Ag)	2	2+6	2+6+10	2+6+10+0	1+0+0+0	0
17	52 (Te)	2	2+6	2+6+10	2+6+10+0	2+4+0+0	0
18	55 (Cs)	2	2+6	2+6+10	2+6+10+0	2+6+0+0	1
19	63 (Eu)	2	2+6	2+6+10	2+6+10+7	2+6+0+0	2
20	72 (Hf)	2	2+6	2+6+10	2+6+10+14	2+6+2+0	2
21	76 (Os)	2	2+6	2+6+10	2+6+10+14	2+6+6+0	2
22	78 (Pt)	2	2+6	2+6+10	2+6+10+14	2+6+9+0	1
23	80 (Hg)	2	2+6	2+6+10	2+6+10+14	2+6+10+0	2

SHRIMAD BHAGWAD GEETA

ELECTRONIC CONFIGURATION TABLES

CHAPTER-1-18

Chapter-1 Text Electronic configuration of silver (Ag 47)									
1	2	3	4	5	6	7	8	9	10
S1	S2	P2	S3	P3	D3	S4	P4	D4	F4
2	2	6	2	6	10	2	6	9	1

Chapter-1 Geeta Parinam Electronic configuration of silver (Ag 47)									
1	2	3	4	5	6	7	8	9	10
S1	S2	P2	S3	P3	D3	S4	P4	D4	F4
2	2	6	2	6	10	2	6	9	1

Chapter-2 Text Electronic configuration (Hf 72)													
1	2	3	4	5	6	7	8	9	10	11	12	13	14
s1	s2	p2	s3	p3	D3	s4	p4	d4	f4	s5	P5	d5	f5
2	2	6	2	6	10	2	6	10	14	2	6	2	2

Chapter-2 Geeta Parinam Electronic configuration (Os 76)													
1	2	3	4	5	6	7	8	9	10	11	12	13	14
s1	s2	p2	s3	p3	D3	s4	p4	d4	f4	s5	P5	d5	f5
2	2	6	2	6	10	2	6	10	14	2	6	6	2

Chapter-3 Text Electronic configuration (Te 43)									
1	2	3	4	5	6	7	8	9	10
s1	s2	p2	s3	p3	d3	s4	p4	d4	f4
2	2	6	2	6	10	2	6	5	2

Chapter-3 Geeta Parinam Electronic configuration (Ag 47)									
1	2	3	4	5	6	7	8	9	10
s1	s2	p2	s3	p3	d3	s4	p4	d4	f4
2	2	6	2	6	10	2	6	10	1

Chapter-4 Text Electronic configuration (Mo 42)									
1	2	3	4	5	6	7	8	9	10
s1	s2	p2	s3	p3	d3	S4	p4	d4	f4
2	2	6	2	6	10	2	6	5	1

Chapter-4 Geeta Parinam Electronic configuration (Rh 45)									
1	2	3	4	5	6	7	8	9	10
s1	s2	p2	s3	p3	d3	S4	p4	d4	f4
2	2	6	2	6	10	2	6	8	1

Chapter-5 Text Electronic configuration (Cu 29)						
1	2	3	4	5	6	7
s1	s2	p2	s3	p3	d3	s4
2	2	6	2	6	10	1

Chapter-5 Geeta Parinam Electronic configuration (Ga 31)							
1	2	3	4	5	6	7	8
s1	s2	p2	s3	p3	d3	S4	p4
2	2	6	2	6	10	2	1

Chapter-6 Text Electronic configuration (Ag 47)									
1	2	3.	4	5	6	7	8	9	10
s1	S2	p2	s3	p3	d3	s4	p4	d4	f4
2	2	6	2	6	10	2	6	10	1

Chapter-6 Geeta Parinam Electronic configuration (Te 52)										
1	2	3	4	5	6	7	8	9	10	11
s1	S2	p2	s3	p3	d3	S4	p4	d4	f4	s5
2	2	6	2	6	10	2	6	10	2	4

Chapter-7 Text Electronic configuration (Zn 28)						
1	2	3	4	5	6	7
s1	s2	p2	s3	p3	d3	S4
2	2	6	2	6	10	2

Chapter-7 Geeta Parinam Electronic configuration (Ga 31)							
1	2	3	4	5	6	7	8
s1	s2	p2	s3	p3	d3	S4	p4
2	2	6	2	6	10	2	1

Chapter-8 Text Electronic configuration (Ni 28)						
1	2	3	4	5	6	7
s1	s2	p2	s3	p3	d3	S4
2	2	6	2	6	8	2

Chapter-8 Geeta Parinam Electronic configuration (ZN 30)						
1	2	3	4	5	6	7
s1	s2	p2	s3	p3	d3	S4
2	2	6	2	6	10	2

Chapter-9 Text Electronic configuration (Se 34)							
1	2	3	4	5	6	7	8
s1	S2	p2	s3	p3	d3	s4	p4
2	2	6	2	6	10	2	4

Chapter-9 Geeta Parinam Electronic configuration (Br 35)							
1	2	3	4	5	6	7	8
s1	S2	p2	s3	p3	d3	S4	p4
2	2	6	2	6	10	2	5

Chapter-10 Text Electronic configuration (Mo 42)									
1	2	3	4	5	6	7	8	9	10
s1	s2	p2	s3	p3	d3	s4	p4	d4	f4
2	2	6	2	6	10	2	6	5	1

Chapter-10 Geeta Parinam Electronic configuration (Rh 45)									
1	2	3	4	5	6	7	8	9	10
s1	s2	p2	s3	p3	d3	S4	p4	d4	f4
2	2	6	2	6	10	2	6	8	1

Chapter-11 Text Electronic configuration (Cs 55)											
1	2	3	4	5	6	7	8	9	10	11	12
s1	s2	p2	s3	p3	d3	S4	p4	D4	f4	s5	p5
2	2	6	2	6	10	2	6	10	2	6	1

Chapter-11 Geeta Parinam Electronic configuration (Eu 63)												
1	2	3	4	5	6	7	8	9	10	11	12	13
s1	s2	p2	s3	p3	d3	S4	p4	d4	f4	s5	P5	d5
2	2	6	2	6	10	2	6	10	2	6	6	2

Chapter-12 Text Electronic configuration (Ca 20)					
1	2	3	4	5	6
s1	s2	p2	s3	p3	D3
2	2	6	2	6	2

Chapter-12 Geeta Parinam Electronic configuration (Ti 22)						
1	2	3	4	5	6	7
S1	S2	p2	s3	p3	d3	S4
2	2	6	2	6	2	2

Chapter-13 Text Electronic configuration (Se 34)							
1	2	3	4	5	6	7	8
s1	S2	P2	S3	p3	d3	s4	p4
2	2	6	2	6	10	2	4

Chapter-13 Geeta Parinam Electronic configuration (Br 35)							
1	2	3	4	5	6	7	8
s1	S2	P2	s3	p3	d3	S4	p4
2	2	6	2	6	10	2	5

Chapter-14 Text Electronic configuration (Co 27)						
1	2	3	4	5	6	7
s1	S2	p2	s3	p3	d3	S4
2	2	6	2	6	7	2

Chapter-14 Geeta Parinam Electronic configuration (ZN 30)						
1	2	3	4	5	6	7
S1	s2	p2	s3	p3	D3	S4
2	2	6	2	6	10	2

Chapter-15 Text Electronic configuration (Ca 20)					
1	2	3	4	5	6
S1	s2	p2	s3	p3	D3
2	2	6	2	6	2

Chapter-15 Geeta Parinam Electronic configuration (Sc 21)						
1	2	3	4	5	6	7
s1	s2	p2	s3	p3	D3	S4
2	2	6	2	6	1	2

Chapter-16 Text Electronic configuration (Cr 24)						
1	2	3	4	5	6	7
s1	S2	p2	S3	p3	D3	S4
2	2	6	2	6	5	1

Chapter-16 Geeta Parinam Electronic configuration (Mn 25)						
1	2	3	4	5	6	7
s1	S2	p2	S3	p3	D3	S4
2	2	6	2	6	5	2

Chapter-17 Text Electronic configuration (Ni 28)						
1	2	3	4	5	6	7
S1	s2	p2	s3	p3	D3	S4
2	2	6	2	6	8	2

Chapter-17 Geeta Parinam Electronic configuration (ZN 30)						
1	2	3	4	5	6	7
s1	s2	p2	s3	p3	D3	S4
2	2	6	2	6	10	2

Chapter-18 Text Electronic configuration (Pt 78)													
1	2	3	4	5	6	7	8	9	10	11	12	13	14
s1	s2	P2	s3	p3	d3	s4	p4	d4	f4	S5	p5	d5	F5
2	2	6	2	6	10	2	6	10	14	2	6	9	1

Chapter-18 Geeta Parinam Electronic configuration (Hg 80)													
1	2	3	4	5	6	7	8	9	10	11	12	13	14
s1	s2	P2	s3	p3	d3	s4	p4	d4	f4	S5	p5	d5	F5
2	2	6	2	6	10	2	6	10	14	2	6	10	2

Chase as Manifestation Layer (3, 4, 5, 6)

Chase of the organization format of *Shrimad Bhagwad Geeta* may be had as manifestation layer (3, 4, 5, 6) with 3-Space in the role of dimension, 4-Space in the role of boundary, 5-Space in the role of domain and 6-Space in the role of origin.

3-Space as *Vyakta,,* our manifest/expressed world, 4-Space as *Avyakta* /un-manifest world, 5-Space as *Avakto-Avyaktat* /base of un-manifest and 6-Space as *Pursha*/existence within human frame, is to cover the existence as *Sathul Sharir*/physical body, *Suksham Sharir*/ Subtle body, *Karan Sharir*/ Casual body and *Divya Pursha* / as of transcendental features.

This is the range of macro world - micro world - nuclear world and base of the nuclear world

This chase is ultimately to cover the range from Triloky to Trinity of Gods. As such this is to be of the order of TRANSCENDENTAL WORLD (5-Space).

*****

Section VI
FEATURES OF BASICS

VEDIC MATHEMATICS FONT

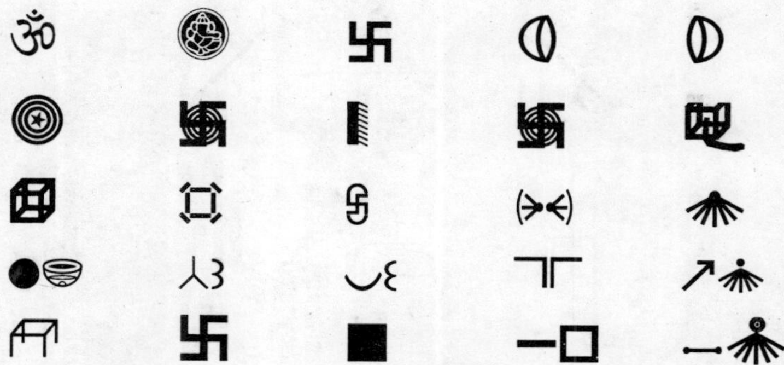

STUDY ZONE

Srimad Bhagwad Gita Study – Zone								
$\frac{a}{2}$	$\frac{a^2}{4a}$	$\frac{a^3}{6a^2}$	$\frac{a^4}{8a^3}$	$\frac{a^5}{10a^4}$ $\frac{a^6}{12a^5}$ $\frac{a^5}{10a^4}$	$\frac{a^2}{8a}$	$\frac{a^3}{6a^2}$	$\frac{a^4}{4a^3}$	
				10x5=50 50x7=350 10x5=50 50x7=350				
2x1 =2	4x2 =8	6x3 =18	8x4 =32		8x4 =32	6x3 =18	4x2 =8	2x1 =2
Orbitals				350+350=700				
2	6	10	14	18=5+6+7				

MEASURING ROD

1-space	*2-space*	*3-space*	*4-space*	*5-space*	*6-space*

FOUNDATION

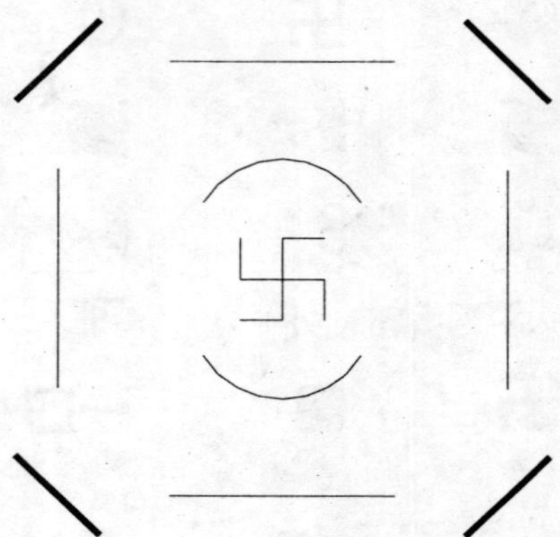

FOUNDATION 11

ELEVEN GEOMETRIES OF 5-SPACE

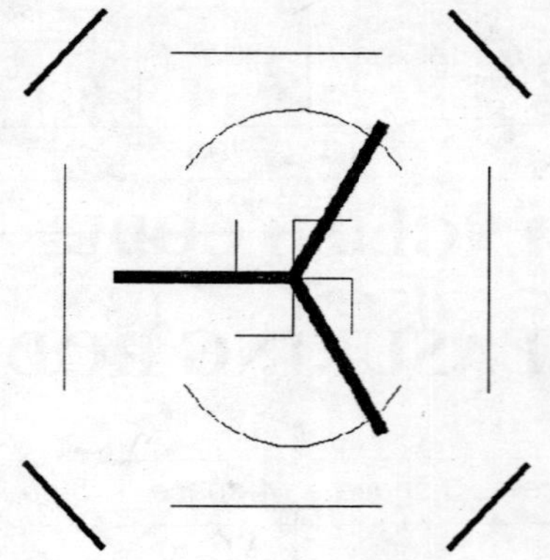

NINE GEOMETRIES OF 4-SPACE

SPACE BIBLE

S	19	19	B	02	02
A	01	20	I	09	11
P	16	36	B	02	13
C	03	39	L	12	25
E	05	44	E	05	30

SPACE BIBLE

44 + 30 = 74

GOD EYE

26+35=6

C H U R C H 03 08 21 18 03 08

61

G E N E R I C07 05 14 05 18 09 03

61

ENGLISH CODE

E	05	05	C	03	03
N	14	19	O	15	18
G	07	26	D	04	22
L	12	38	E	05	27
I	09	47	(74 + 27 = 101)		
S	19	66			
H	08	74			

ULTIMATE
101

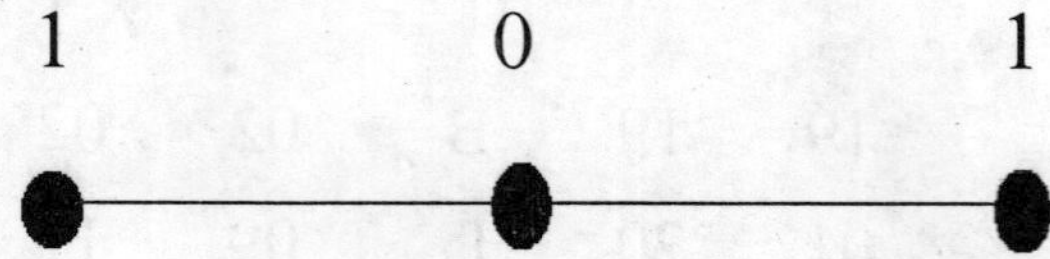

ULTIMATE INTERVAL 101
DIVISON
101

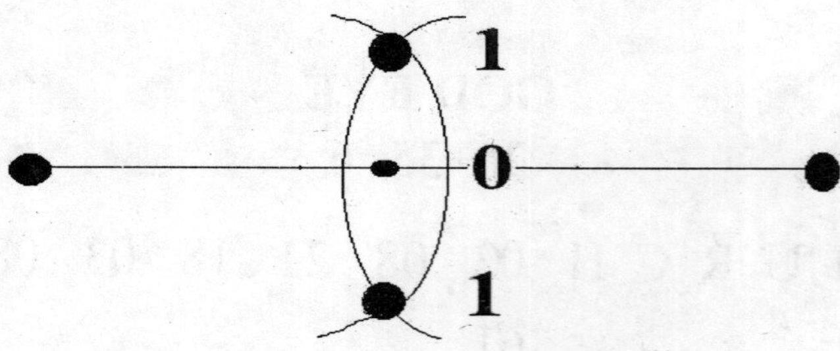

ENGLISH CODE
74, 27

PAiRING HALF
74, 27 HALF
13+1+13
GOD GRACE
26+ONE

GRACE FREE

ONE

G	07	07	F	06	06
R	18	25	R	18	24
A	01	26	E	05	29
C	03	29	E	05	34
E	05	34			

ONE

15 14 05

34

ENGLISH PAIRING

A = 01	N = 14
B = 02	O = 15
C = 03	P = 16
D = 04	Q = 17
E = 05	R = 18
F = 06	S = 19
G = 07	T = 20
H = 08	U = 21
I = 09	V = 22
J = 10	W = 23
K = 11	X = 24
L = 12	Y = 25
M = 13	Z = 26

SUN GOD

54, 26

CREATOR

C	03	03	
R	18	21	
E	05	26	GOD
A	01	01	
T	20	21	
O	15	36	
R	18	54	SUN

GOD LABORATORY

G	07	07	L	12	12
O	15	22	A	01	13
D	04	26	B	02	15
GOD		O	15	30	
26		R	18	48	
LORD		A	01	49	
49		T	20	20	
			O	15	35
			R	18	43
AMBROSIA		Y	25	78	
78				GOD	LORD
AMBROSIA					
26	26+23		26+23+29		

FOUR-HEAD LORD

127

FOUR-HEAD LORD

60+18 49 127

LABORATORY

127

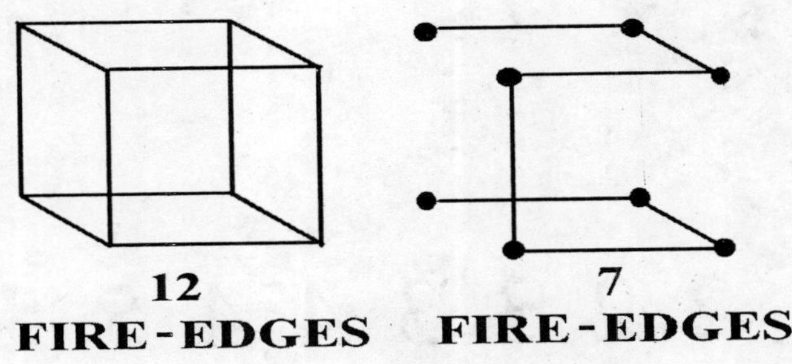

GOD LABORATORY

01	02	03	04	05	06	07	08	09
10	11	12	13	14	15	16	17	18
19	20	21	22	23	24	25	26	27
28	29	30	31	32	33	34	35	36
37	38	39	40	41	42	43	44	45
46	47	48	49	50	51	52	53	54
55	56	57	58	59	60	61	62	63
64	65	66	67	68	69	70	71	72
73	74	75	76	77	78	79	80	81
82	83	84	85	86	87	88	89	90
91	92	93	94	95	96	97	98	99

	01	02	03	04	05	06	07	08	09
	10	11	12	13	14	15	16	17	18
		20	21	22	23	24	25	26	27
23			30	31	32	33	34	35	36
				40	41	42	43	44	45
+					50	51	52	53	54
6						60	61	62	63
							70	71	72
=								80	81
29									90
=									
	1	2	2	3	3	4	4	5	5

BIBLE GENERIC

NVF 6

ABBA

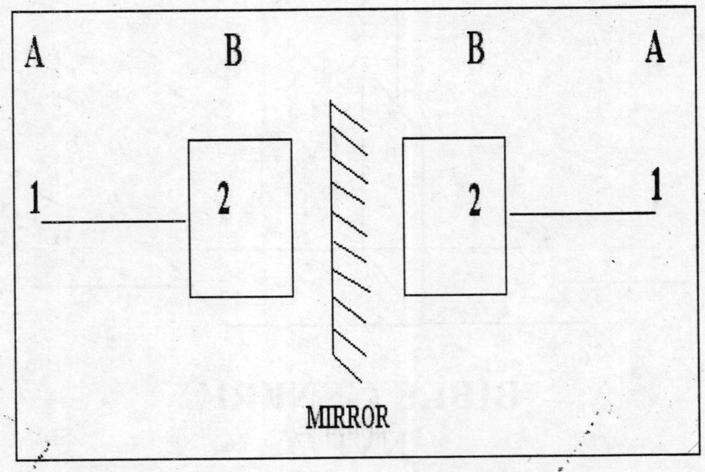

NVF (BIBLE GENERIC) = 91

NVF (MIRROR) = 91

BIBLE GENERIC

NVF 6

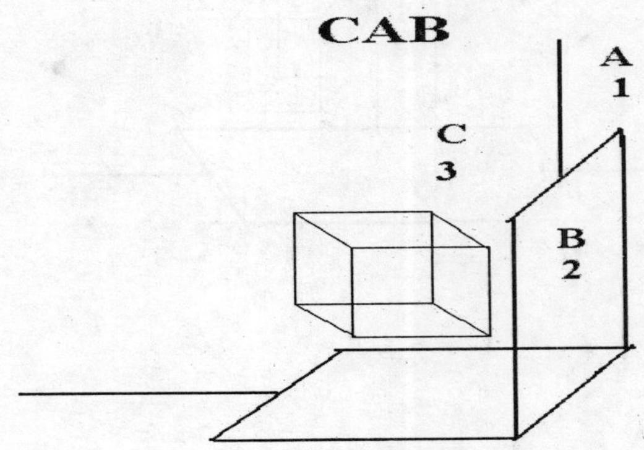

BIBLE GENERIC

NVF 7

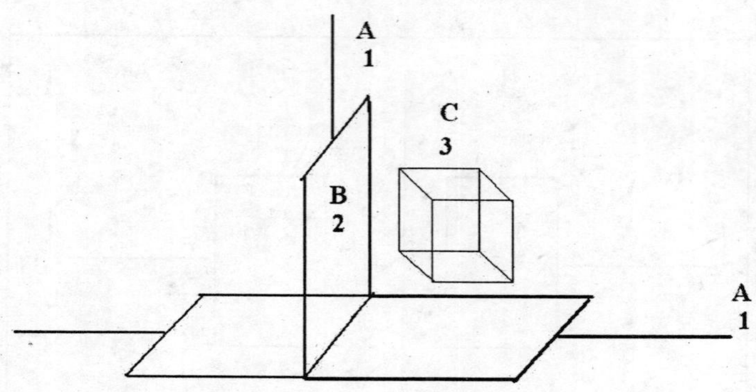

BIBLE GENERIC

NVF 7

BAD

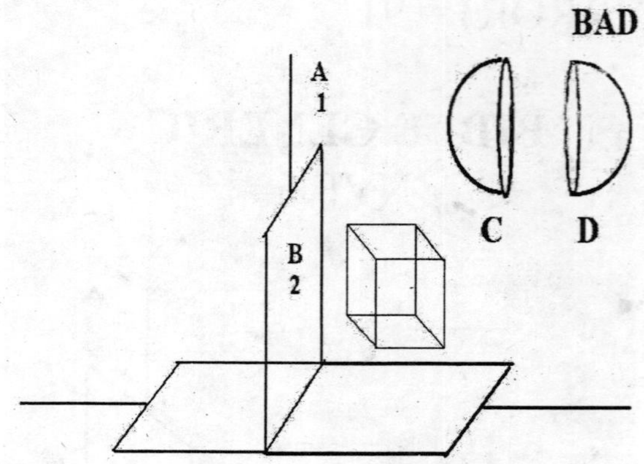

BIBLE GENERIC

NVF 13

ABIA

13 EDGED CUBE

1, 2, 9, 1

(1)	A	axis	1
(2)	B	base	2
(3)	I	Origin	9
(4)	A	13th Edge	1

BIBLE GENERIC

NVF 13

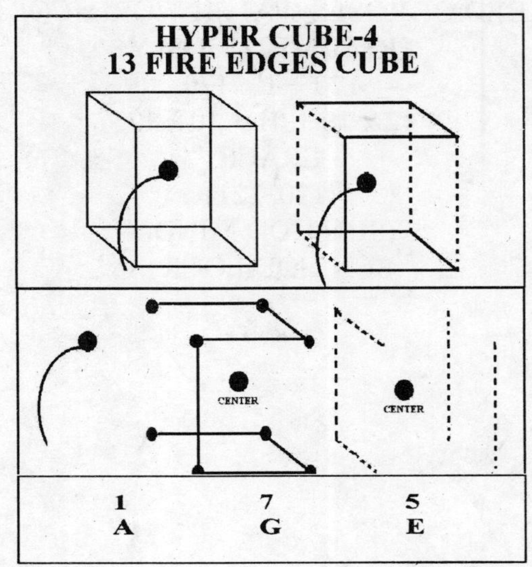

BIBLE GENERIC

NVF 14

ABIB

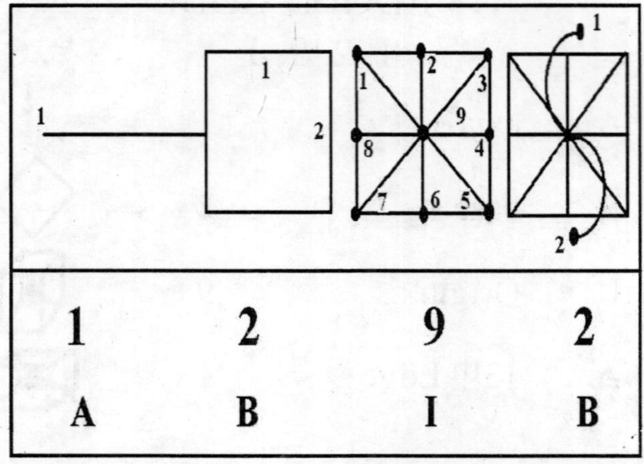

BIBLE GENERIC

NVF 241

A BED OF ROSES WATCH WORD: CODE
PRESUMPTUOUSLY
120 + 1 + 120
12 x 10 + 01 + 10 x 12
GRACE
(120, 121)
(SUN GOD MIND,
SUN GOD LOCK)

******

Section VII

INITIAL LESSONS

Lesson 1

Ganita Sutras

Ganita Sutras include Up-Sutras. There are 16 Sutras and 13 Up-Sutras. The entire range of Ganita Sutras text of 29 Sutras and Up-Sutras avails in all only 520 letters. As such, this text deserves to be memorized for its convenient reference and applications. Here in this lesson is reproduced text of 16 Sutras while 13 Up-Sutras text is taken up in the following lesson:

Sutra No.	Text in Sanskrit
1	ॐ । एकाधिकेन पूर्वेण
2	निखिलं नवतश्चरमं दशतः
3	ऊर्ध्वतिर्यग्भ्याम्
4	परावर्त्य योजयेत्
5.	शून्यं साम्यसमुच्चये
6.	(आनूरूप्ये) शून्यमन्यत्
7.	संकलनव्यवकलनाभ्याम्
8.	पूरणापूरणाभ्याम्
9.	चलनकलनाभ्याम्
10.	यावदूनम्
11.	व्यष्टिसमष्टिः
12.	शेषण्यङ्.केन चरमेण
13.	सेपान्त्यद्वयमन्त्यम्
14.	एकन्यूनेन पूर्वेण
15.	गुणितसमुच्चयः
16.	गुणकसमुच्चयः

For facility of those who are not well acquainted with Devnagri Script, here is reproduced the text in Devnagri Script together with Roman Script:

No.	Devnagri Script	Roman Script
1.	ॐ। एकाधिकेन पूर्वेण	Om.Ekadhikena Purven
2.	निखिलं नवतश्चरमं दशतः	Nikhilam Navata scaramam Daosatah
3.	ऊर्ध्वतिर्यग्भ्याम्	UrdhvaTiryagbhyam
4.	परावर्त्य योजयेत्	Paravartya Yojayet
5.	शून्यं साम्यसमुच्चये	Sunyam Samyasamuccaye
6.	(आनूरूप्ये) शून्यमन्यत्	(Anurupye) sunyamanyat
7.	संकलनव्यवकलनाभ्याम्	Sankalana Vyavakalanbhyam)
8.	पूरणापूरणाभ्याम्	Puranapuranabhyam
9.	चलनकलनाभ्याम्	Calana Kalanabhyam
10.	यावदूनम्	Yavadunam
11.	व्यष्टिसमष्टिः	Vyastisamasihti
12.	शेषण्यङ्.केन चरमेण	Sesanyankna Caramena
13.	सेपान्त्यद्वयमन्त्यम्	Sopantyadvayamantyam
14.	एकन्यूनेन पूर्वेण	Ekanyunena Purveṇa
15.	गुणितसमुच्चयः	Gunitsamuccayah
16.	गुणकसमुच्चयः	Gunakasamuccayah

*****

Lesson 2

Ganita Upsutras

Here follows the text of 13 Ganita Upsutras:

Sutra No.	Text in Sanskrit
1.	प्रणवः । आनुरूप्येण
2.	शिष्यते शेषसंज्ञः
3.	आद्यमाद्येनान्त्यमन्त्येन
4.	केवलैः सप्तकं गुण्यात्
5.	वेष्टनम्
6.	यावदूनं तावदूनम
7.	यावदूनं तावदूनीकृत्य वर्गं च याजयेत्
8.	अन्त्ययोर्दशकेऽपि
9.	अन्त्ययोरेव
10.	समुच्चयगुणितः
11.	लोपनस्थापनाभ्याम्
12.	विलोकनम्
13.	गुणितसमुच्चयः समुच्चयगुणितः

For facility of those who are not well acquainted with Devnagri Script, here is reproduced the text in Devnagri Script together with Roman Script:

No.	Devnagri Script	Roman Script
1	प्रणवः । आनुरूप्येण	Parnavah. Anurupyena
2	शिष्यते शेषसंज्ञः	Sisyate Sesasmjnah
3	आद्यमाद्येनान्त्यमन्त्येन	Adyamadyenan-tyamantyena
4	केवलैः सप्तकं गुण्यात्	Kevalaih Saptakam Gunyat
5.	वेष्टनम्	Vestanam
6.	यावदूनं तावदूनम	Yavadunam Tavadunam
7.	यावदूनं तावदूनीकृत्य वर्गं च याजयेत्	Yavadunam Tavadunikrtya Vargam ca Yojget
8.	अन्त्ययोर्दशकेऽपि	Antyayordasakepi
9.	अन्त्ययोरेव	Antyayoreva
10.	समुच्चयगुणितः	Samuccayagunitah
11.	लोपनस्थापनाभ्याम्	Lopanasthapanabhyam
12.	विलोकनम्	Vilokanam
13.	गुणितसमुच्चयः समुच्चयगुणितः	Gunitasamuccayah Samuccayagunitah

*****

Lesson 3

Ganita Sutra-1

एकाधिकेन पूर्वेण
Ekadhiken Purvena
One more than before

1. Full text of Sutra-1 is एकाधिकेन पूर्वेण (Ekadhiken Purvena). Its simple English rendering as a working rule, comes to be, "One more than before". Total range of text of the Sutra is of 16 letters:

1	2	3	4	5	6	7	8	9	10	11	12	13	14	15	16
ए	क्	आ	ध्	इ	क्	ए	न्	अ	प्	ऊ	र्	व्	ए	ण्	अ

2. This text avails two words
   (i) एकाधिकेन Ekadhiken
   (ii) पूर्वेण: Purvena

3. The first word is the first conceptual term. This admits sub formulations as
   (a) एकाधिकेन Ekadhiken
   (b) एक+अधिकेन Ek+adhiken
   (c) अधि+केन adhi+ken

एकाधिकेन	एक	अधि	केन
Ekadhiken	Ek	Adhi	Ken

4. Simple English rendering for sub-formulations may be as

एकाधिकेन	एक	अधि	केन
Ekadhiken	Ek	Adhi	Ken
One more than of which	One	More	Than of which

5. The three sub-formulations and these together as bigger formulation, which in a way expects a sort of answer for the question, which is expected from the second part of the text, that is Purvena.

6. The conceptual term as एकाधिकेन Ekadhiken is, it deserves to be understood in terms of its three sub-formulations.

First Sub formulation:

एक	Eka	One

7. In Vedic systems, one is whole some source concept of the order and form as ओम् एकअक्षर ब्रह्म (Om Ekakshar Braham).

8. The composition of formulation एक (Eka) is of one vowel ए and consonant क्.

9. The vowel ए is the sixth vowel

10. The consonant क् is the first varga consonant. The scriptures enlighten about this as Chaturmukhi (four head Lord, Brahma, creator the supreme).

11. Lord Brahma is the overlord of 4-Space and Lord Vishnu is the overlord of 6-Space.

12. Sathapatya Up Ved (in particular Mansara) enlightens about the measuring rod as well as the measure as that Lord Vishnu is the Lord of the measuring rod and Lord Brahma is the lord of measure.

13. Further, 4-Space plays the role of dimension of 6-Space and as such the measure for Vishnu Lok is supplied by creator the supreme, Lord Brahma himself.

14. This way the conceptual formulation एक (Eka/one) is the unit hypercube-6 as representative regular body of 6-Space presided by Lord Vishnu.

15. Accordingly the Ganita Sutras are organized for the Mathematics, Science and technology of 6-Space.

Second Sub-formulation:

अधि	Adhi	More

16. The formulation अधि (Adhi/More) is Upsarga.
17. There are twenty-two upsarga.
18. Sarga means creation and upsarga means localized creations.
19. Upsargas concentrate within the given domains of creations.
20. Adhi, as upsarga, in the nature of the upsarga, makes the domain of operation of ONE (Ek) to be Ekadhi (one more) and this upsarga (adhi), in the nature of its role, also gets extended to the following sub-formulation केन (ken), which literally is a formulation to raise question, and this as such makes the role of this formulation (adhi) as to make the application of the nature of an EXPECTING ONE.

Third Sub-formulation:

केन	Ken	Question: than of which	Who?

21. There is Kena-Upanishad; the Upanishad which poses and answers question about Braham. The first mantra of this Upanishad poses the basic questions

ओम् केनेषितं पतति प्रेषितं मनः
केन प्राणः प्रथमः प्रैति युक्तः ।
केनेषितां वाचमिमां वदन्ति
चक्षुः श्रोत्रं क उ देवो युनक्ति ।।१।।

Willed by whom does the directed mind go towards its object? Being directed by whom does the vital force that precedes all, proceed (towards its duty)? By whom is this speech willed that people utter? Who is the effulgent being who directs the eye and the ears?

The commentary and the meanings of the Mantra are approached as:

Kena, by what agent; being is it am, willed, directed;.........

22. Shrimad Bhagwad Geeta in its Shaloka 34 of Chapter-4, makes it clear about the mode of getting knowledge by posing questions. The Shaloka enlightens about this aspect as:

    Original text
    तद् विद्धि प्रणिपातेन परिप्रश्नेन सेवया।
    उपदेक्ष्यन्ति ते ज्ञानं ज्ञानिनस्तत्त्वदर्शिनः।।३४।।

    Text in Roman Script
    tad viddhi pranipatena pariprasnena sevaya
    upadeksyanti te jnanam jnaninas tattva-darsinah

    Meanings rendered by Maharishi Mahesh Yogi:
    Know this: through homage, repeated inquiry and service, the men of knowledge who have experienced Reality will teach you knowledge.

23. As such the formulation केन Ken as part of very first conceptual term Ekadhikena, as such is the basic foundation pillar of the text.

24. The conclusion statement in the context may be that this conceptual term as such Adhi (more) as upsarga at the middle bridges by expecting from the formulations on its either sides and finally the formulation on the whole keeps expecting the following formulation as an answer for the inbuilt question and it is only with the answer following from the following part of the text as second conceptual term that is purvena which as simple rendering to mean the before.

*****

Lesson 4
Number Cone

Introductory

There is a common pool of Ancient Wisdom with Ved as the core source thereof. Vedic knowledge traditions are to urge to know Ved in terms of Ved itself. The organized knowledge of Ved as four Vedas, Rig, Yaju, Sam and Atharav accepts knowing first three Vedas in terms of fourth Ved, Atharav Ved. The first quarter of first Mantra of Atharav Ved enlightens about the first principle of knowing Vedic knowledge. The Mantra, as its first quarter enlightens:

Yeh Trishapta Paryanti Vishwa

ये त्रिशप्ता पर्यन्ति विश्वः

This Trishapta envelops Vishwa.

"Trishapta/ त्रिशप्ता" is a technical term, whose one prominent expression in terms of artifices of numbers is "3 and 7"; of which, the artifices coordinations organizations, come to be 37 and 73. There are other coordinations organizations of artifices of 3 and 7 as well like 3 + 7, 3/7, 7/3, 3 x 7, 3^7, 7^3, 3.7, 7.3 and so on. However, here at this initial stage of very first lesson of this introductory course of learning and teaching simultaneously for students and teachers together, the initial focus at initiation stage of Discipline of Mathematics, in terms of Numbers as tools as counts and formats for coordination and organization of artifices of numbers, the initial coordinations organization of artifices of 3 and 7 are being focused as 37 and 73.

This coordination organization focus of artifices of 3 and 7 as 37 and 73 is availed for organization of knowledge of "Sidha Sidhanta Padati" also known as Gorakshko Upanishad, the source scripture of Shiva cult sect, Gorakh Panthi. The scripture is of six Updesha (enlightenment sermons). The first Updesha is titled "Pinda Utpati/ emergence of body within human frame from within another body within human frame". The range of this Updesha is of 73 shalokas length. The middle shaloka, 37 shaloka, enlightens about Asth Sakar Murti/hypercube-4, representative regular body of creator's Space/4-Space enveloped within eight solid boundary components.

The enlightenment up till sixth Updesha is of the transcendental order of transcendence into Pursha domain/orb of Sun/Vishnu Lok/ 6-Space. Further Upanishadic enlightenment is that the Vishnu Lok/ 6-Space/Pursha format accepts 26 tatav/basic elements. Sankhya approach to these basic elements is along the formats of the artifices of numbers 1 to 26. Initially it may sound all strange if stated as that English Alphabet of 26 letters (A to Z) talks mathematics of the coordinations organizations of artifices of numbers 1 to 26 parallel to letters A to Z but the studies of the author convince as that this is there because of fundamental unity of Ancient wisdom identically feeding from same course source all the frequencies which are accepted by human mind from generation to generation.

Pinda Utpati/emergence of body, is conceptually, about the phenomena of birth of a child. The number value format (NVF) for CHILD for the letters availed (C = 3, H = 8, I = 9, L = 12 and D = 4 total 36) and number value format (NVF) for CHILDREN for the letters availed (CHILD = 36, R = 18, E = 5, and N = 14 total 73), as such takes us to the identical counts formats for the organizations of knowledge.

NVF (NUMBER) = NVF (COUNT) = NVF (FORMAT) = 73 and reflection pair of 73 is 37 and same is NVF (CONE). As such the title of the first lesson is settled as NUMBER CONE.

Number cone

Number as count, and as format, as artifice of 73 accepting organization as 36 +1 + 36, and further as 6 x 6 + 1 x 1 + 6 x 6 takes us to the spatial/plane/square/ 2-Space order. The only aspect being availed here for the initial stage of approaching numbers as counts is that the pairing with NVF (PAIRING) = 74 = 37 + 37, is being had such that we to have a pair of cones and then to proceed for coordination organization for the counts in terms of pair of boundary lines of spatial cut of a cone as under:

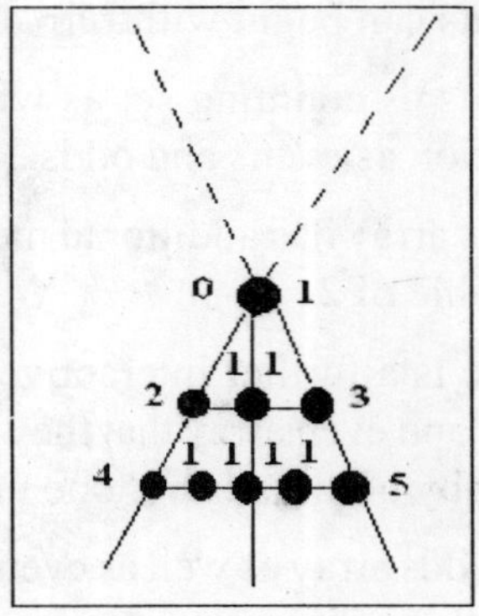

Lesson Steps:

Stage-1: To help the students to draw above number-cone.

Stage-2: To help the students to familiarize with the depiction and placements of points as well as fixations of linear units along different steps of cone within its pair of boundary lines as spatial/plane/2-Space print out of the cone as above.

Stage-3: To specifically chase points counts and linear units count as

1. One point is one count.
2. Three points are three counts. Three points also fix a pair of units (linear units as counts).
3. Five points are five counts. Five points also fix quadruple units (linear four counts fixed by five points).
4. In general N points help fix (n-1) linear units

Stage-4: To teach counting chase along above number–cone format.

Stage-5: Point out as that in this exposure of this approach to whole numbers set as counting numbers set special features of organization of counting numbers as whole numbers being focused are:

1. As that counting and table of 2 are being simultaneously formatted.

2. Counts can begin with zero as well as with one.
3. Counts as counting set as whole numbers accepts partition as evens and odds.
4. Evens array has additional message as that it is also the table of 2.
5. There is a further interactive message of the odds array and even array that they can interact with each other by both rules of "one more" and "one less".
6. The odds array as well as evens array further convey that these are numbers lines with one jump in between its consecutive members placements.
7. The odds array can also give us 1, 1 x 3, 1 x 5…. While evens array can give us 2, 2 x 4, 2 x 4 x 6….

Note: The following stage-6 is being added just to feed the curiosity as to why the above seventh feature is being mentioned; the focus may be that like partition of whole numbers as odds and evens, the dimensional spaces as well accept parallel partition as odd dimensional spaces and even dimensional spaces and that ultimately the artifices of numbers are to run parallel to the dimensional frames of spaces.

Stage-6: It may be introduced as that odds array supply 1, 1 x 3, 1 x 5…. Shall be taking us to dimensional order ranges of odd dimensional spaces while the evens array supply 2, 2 x 4, 2 x 4 x 6…. shall be taking us to two dimensional order ranges of even dimensional spaces

*****

Lesson 5

Domain Boundary Ratio

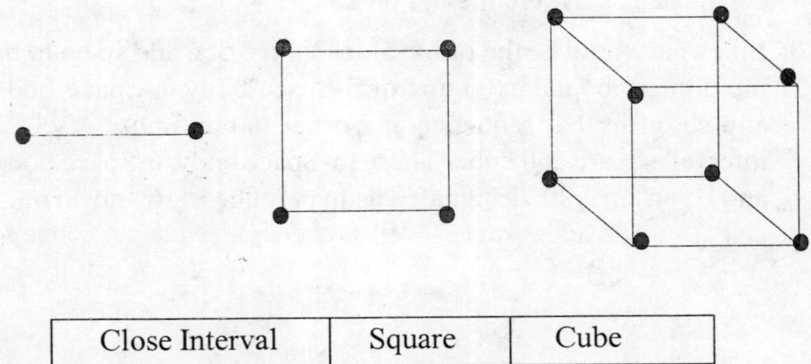

Close Interval	Square	Cube

1. Close interval gives us domain (length) as A^1 and the boundary (pair of end points) as $2A^0$.
2. This becomes the ratio $A^1 : 2A^0$.
3. Square gives us domain (area) as A^2 and the boundary as $4A^1$.
4. This becomes the ratio $A^2 : 4A^1$.
5. If we have a close look at the ratios

   (i) $A^1 : 2A^0$ and

   (ii) $A^2 : 4A^1$,

   we can decipher the pattern for these ratios as

   $A^n : 2nA^{n-1}$ for n = 1 and 2.
6. Now if we put the value n = 3, we get the ratio

   (iii) $A^3 : 6A^2$.
7. This is precisely domain (volume) of cube as A^3 and boundary (six surfaces) of cube as $6A^2$.
8. Once the ratio $A^n: 2nA^{n-1}$ holds for n = 1, 2 and 3 for our known geometric bodies, close interval, square and cube as representative regular bodies of 1-space, 2-space and 3-space respectively, the query would be as to how this sequence for values n = 4, 5, 6 and for onward values would behave?

9. Natural answer would be that the ratio values for n = 4 is to be the domain boundary ratio of 4-Space body in continuity of close interval, square and cube. It (4-Space body) may be designated as hypercube-4.

10. Likewise would be the ratio values for n = 5, 6 and so on to be the domain boundary ratios of 5-Space body, 6-Space body and so on in that sequence and order in continuity of close interval, square and cube. These (5-Space body, 6-Space body and so on) may be designated as hypercubes 5, 6 and so on.

*****

Lesson 6

Geometric Components Formulation of Interval, Square and Cube

$(A+2)^n$, n=1,2,3

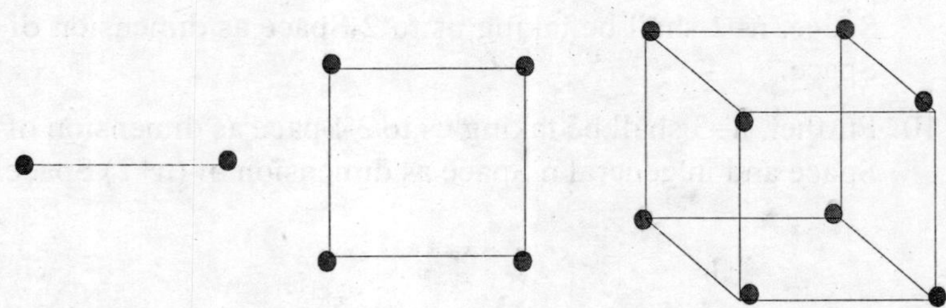

1. The arithmetic expression $(A+2)^n$ for n=1 can be re-expressed as:

$$(A^1 + 2A^0)^1.$$

2. This expression gives us parallel geometric entities as A^1 as length of interval and as $2A^0$ as a pair of end points of a close interval.

3. The arithmetic expression $(A+2)^n$ for n=2 can be re-expressed as:

$$(A^1 + 2A^0)^2 = (A^2 + 4A^1 + 4A^0).$$

4. This expression gives us parallel geometric entities as A^2 as area of a square and as $4A^1$ as four boundary lines of square and as $4A^0$ as four corner points of a square.

5. The arithmetic expression $(A+2)^n$ for n=3 can be re-expressed as:

$$(A^1 + 2A^0)^3 = (A^3 + 6A^2 + 12A^1 + 8A^0).$$

6. This expression gives us parallel geometric entities as A^3 as volume of a cube and as $6A^2$ as six surface plates of a cube, as $12A^1$ as twelve edges of a cube and $8A^0$ as eight corner points of a cube.

7. The above property of need of adding two units to length A is a property which would help us to have comparative relationship of 1-space as dimension of (1+2) space.

8. In general, the above property of need of adding two units to the value A is a property which would help us to have comparative relationship of n-Space as dimension of (n+2) Space.

9. In particular while n=1 leads to 1-Space as dimension of 3-Space, n=2 shall be taking us to 2-Space as dimension of 4-Space.

10. Further, n=3 shall be taking us to 3-Space as dimension of 5-Space and in general n Space as dimension of (n+2) Space.

*****

Lesson 7
Existence of Higher Spaces

1. We can view
   (a) interval (line) / 1-space as a track of a moving point,
   (b) surface (square) / 2-space as a track of a moving interval / line / 1-space, and,
   (c) solid space (cube) / 3-space as a track of a moving surface / square / 2-space.
2. From here we can pose a question to ourselves as to what would be the track of a moving solid / cube / 3-space?
3. The natural answer is 4-space / hypercube-4 / hyper solid domain.
4. We are having free movements of solids. Even when we move, we move as a solid block, so we always create 4-space for our free motion.
5. So our space in which we move freely is not a 3-space. It is 4-space and higher spaces all compactified.
6. With the motion of hypercube-4 / 4-space, we shall be having 5-space for us and in general with the motion of hypercube-n / n-space, we shall be having (n+1) space.
7. As such the existence of higher spaces is one concept for whose comprehension and chase one shall perfect one's intelligence beginning with the motions of point, line, square and cube.
8. Once one has complete comprehension of existence of 4-Space with hypercube-4 as its representative regular body with domain boundary ratio as $A^4:8B^3$, one shall proceed further for comprehension of 5-Space and its representative regular body.
9. Once one has complete comprehension of existence of 5-Space with hypercube-5 as its representative regular body with domain boundary ratio as $A^5:10B^4$, one shall proceed further

for comprehension of 6-Space and its representative regular body.

10. Initially, for perfection of intelligence, one shall sequentially go on chasing dimensional spaces 1 to 7 and their representative regular bodies. The subsequent chase of higher spaces shall be taken up only after one has perfected one's intelligence in respect of 1 to 7 spaces.

*****

Lesson 8

Outward and Inward Expansions

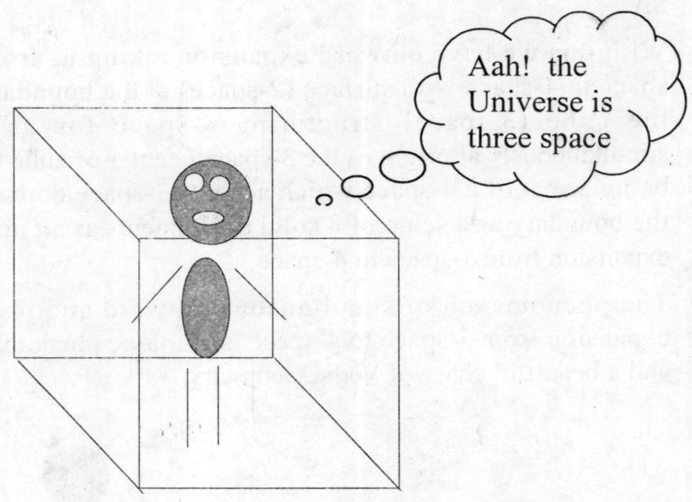

1. 4-space as a track of moving solid (cube), helps us to pause at the static position of a cube as within 4-space.
2. As the cube is wrapped within surfaces, so we can comprehend the situation as that 4-space is enveloping 3-space in terms of surfaces (2-space).
3. The surface plate of a cube when would be stripped off would fall within 4-space.
4. We can see conditioning and structuring of 4-space in terms of the surfaces (2-space).
5. This is 2-space playing the role of dimension of 4-space.
6. Cube is so structured that it at its center permits split up as 8 cubes parallel to the Cartesian cut of 3-space of 8 octants.
7. We can chase enveloping of the center of a cube as 8 cubes coming closer and closer.

8. This is as if the space at the center of cube / origin of Cartesian dimensional frame of 3-space / origin of 3-space is admitting wrapping within solid boundary of 8 components; a 4-space setup wrapped within solid boundary of 8 components in unison with the domain-boundary ratio A^4 : $8A^3$.
9. With this we have outward expansion taking us from 3-space to 4-space with surface (2-space) at the boundary of the cube (3-space) structuring 4-space for us and simultaneously at origin of the 3-space / center of cube there being a seat of a 4-space which accepts 3-space domain at the boundary as a setup of 8 solid components as an inward expansion from 3-space to 4-space.
10. The phenomenon of simultaneous outward and inward expansion from 3-space to 4-space is a unique phenomenon and a beautiful chase of Vedic Geometry.

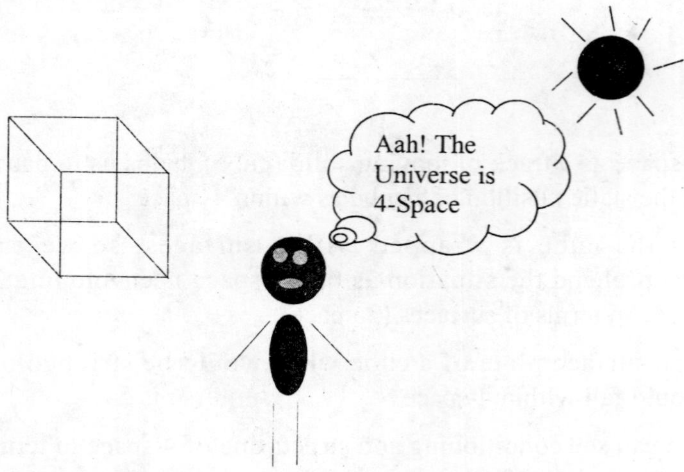

11. The diagram at the outset of this lesson above and the following diagram, when chased together as depictions of situations as to how, as is the situation of diagram-1 above, we first condition ourselves by accepting ourselves as if we are within 3-space and then starting working out linear orders and getting sealed at 3-space.

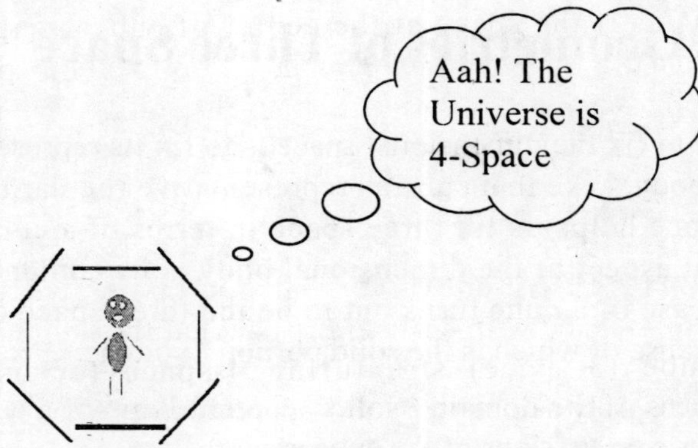

12. If we remind ourselves that we can come out of it and actually come out of it and then the change from the state of first diagram to the second diagram would amount to convincing ourselves that we are in a 4-Space. So perfected intelligence, further shall be availed for taking ourselves out of initial position of being within hypercube-4 and then coming out of it to be in 5-Space and this chase may be continued farther and farther:

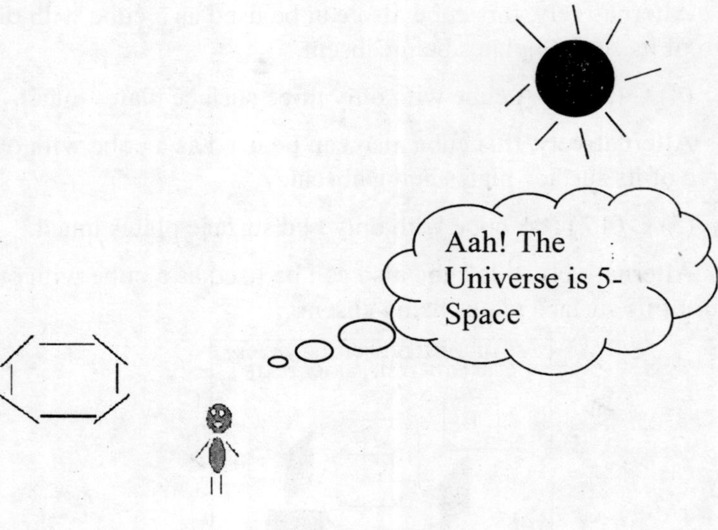

*****

Lesson 9

Geometries of Three Space

One way to fix the dimensional space is to fix its representative regular body. Like that cube as representative regular body of three space helps us fix three space in terms of a cube. The dominant aspect of the dimensional body is its content lump. This in case of a cube turns out to be the three space content lump because of which is the solid portion / volume.

In terms of the domain / solid / content lump of a cube, we can have seven versions of the cube as:

(1) C (0,6) : A cube with all the six surface plates intact.

Alternatively, this cube also can be used as a cube with none of its surface plates being absent.

(2) C (1,5) : A cube with only five surface plates intact.

Alternatively, this cube also can be used as a cube with only one of its surface plates being absent.

(3) C (2,4) : A cube with only four surface plates intact.

Alternatively, this cube also can be used as a cube with only two of its surface plates being absent.

(4) C (3,3) : A cube with only three surface plates intact.

Alternatively, this cube also can be used as a cube with only three of its surface plates being absent.

(5) C (4,2) : A cube with only two surface plates intact.

Alternatively, this cube also can be used as a cube with only four of its surface plates being absent.

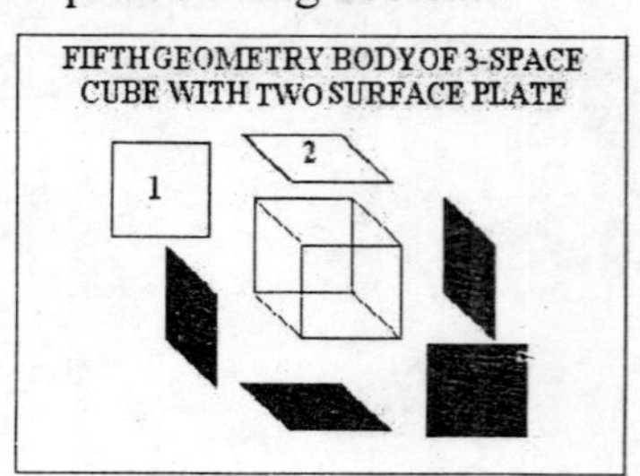

(6) C (5,1) : A cube with only one surface plate intact.

Alternatively, this cube also can be used as a cube with only five of its surface plates being absent.

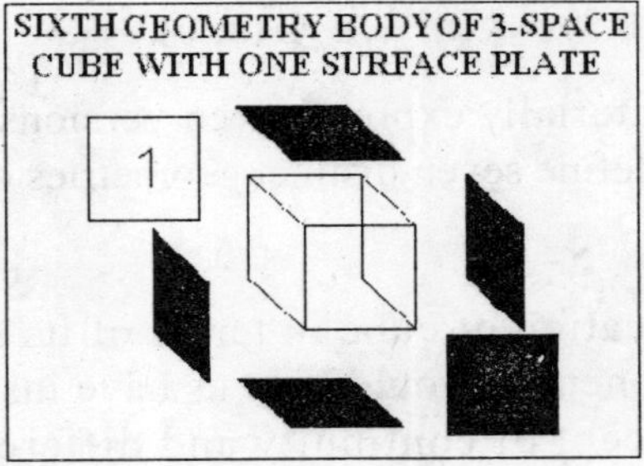

(7) C (6,0) : A cube with all surface plates absent.

Alternatively, this cube also can be used as a cube with none of its surface plates being present.

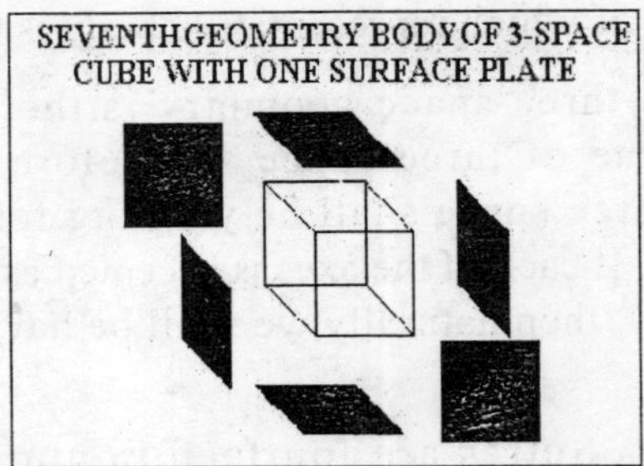

Above seven versions of the cube because of their distinct organization make them geometrically distinct and it is this distinction which makes their geometries distinct. As such we can say that three space admits seven distinct geometries.

We can tabulate as under:

Sr. No.	Geometry of three space	Representative body of three space
1	G (0, 3)	C (0, 6)
2	G (1, 3)	C (1, 5)
3	G (2, 3)	C (2, 4)

4	G (3,3)	C (3, 3)
5	G (4,3)	C (4, 2)
6	G (5,3)	C (5, 1)
7	G (6,3)	C (6, 0)

We can easily pictorially express seven versions of cube. Also we can formally define seven distinct geometries of three space.

Note:

[1] This classification of cube in terms of its surface plates (boundary components), would help us have insight about our present day concept of continuity and differentiation. The continuity is there because of the domain and differentiation is there because of the boundary. Once we ignore this distinction, we stand confronted with a situation which takes us to everywhere continuous but nowhere differentiable functions which in the context would be there because of seventh geometry G(6,3) of which representative body is C(6,0).

[2] Each of the three space geometry is there within three dimensions frame of three space. Therefore, all the seven geometries of three space shall be yielding for us $3 \times 7 = 21$ dimensional axes. If each of the axes is accepted as of two parts (of di-monad format), then naturally we shall be having $21 \times 2 = 42$ components.

[3] Maheshwara Sutras are fourteen in number and these together coordinate 42 letters of Devanagari alphabet.

*****

Lesson 10

2n+1 Geometries for n space

Interval as representative body of 1-Space

The domain boundary ratio of Interval accepts formulation $A^1{:}2B^0$

As such Interval has three versions, namely, close interval, half-close interval, and open interval.

These three distinct versions of interval are representative bodies of 3 geometries of 1-space.

Square as representative body of 2-Space

The domain boundary ratio of Square accepts formulation $A^2{:}4B^1$

As such Square has five versions, namely, Square with all boundary lines, with three boundary lines, with two boundary lines, with one boundary line, and with no boundary line.

These five distinct versions of Square are representative bodies of 5 geometries of 2-space.

Cube as representative body of 3-Space

The domain boundary ratio of Cube accepts formulation $A^3{:}6B^2$

As such Cube has seven versions, namely, Cube with all the surface plates, and with five, four, three, two, one and no surface plate.

These seven distinct versions of Cube are representative bodies of 7 geometries of 3-space.

Hyper Cube-4 as representative body of 4-Space

The domain boundary ratio of Hyper cube-4 accepts formulation $A^4{:}8B^3$

As such Hyper Cube-4 has nine versions, namely, Hyper Cube-4 with all the eight solid boundary components, and with seven, six, five, four, three, two, one and no solid boundary component.

These nine versions of Hyper Cube-4 are representative bodies of 9 geometries of 4-space.

FIRST GEOMETRY	
SECOND GEOMETRY	THIRD GEOMETRY

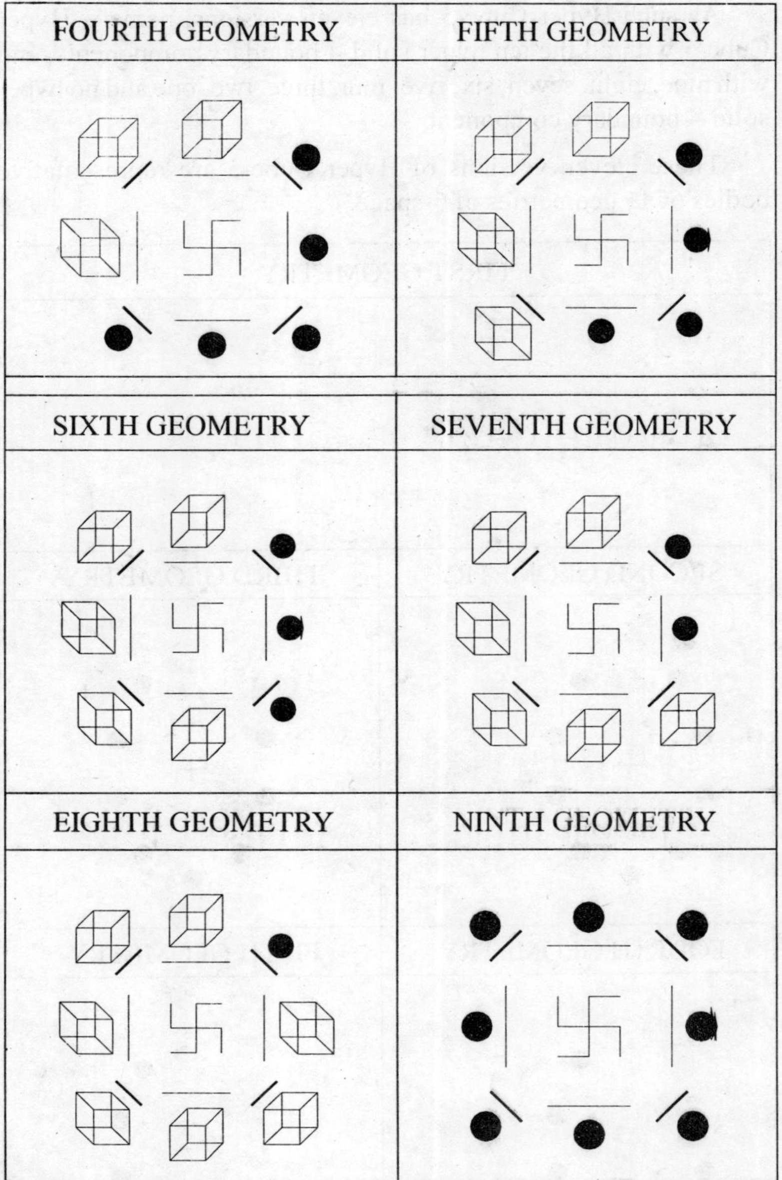

Hyper Cube-5 as representative body of 5-Space

The domain boundary ratio of Hyper cube-5 accepts formulation $A^5:10B^4$

As such Hyper Cube-5 has eleven versions, namely, Hyper Cube-5 with all the ten hyper solid-4 boundary components, and with nine, eight, seven, six, five, four, three, two, one and no hyper solid-4 boundary component.

These eleven versions of Hyper Cube-5 are representative bodies of 11 geometries of 5-space.

FIRST GEOMETRY	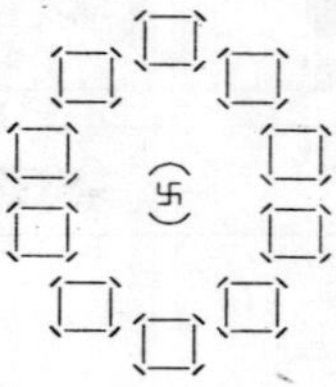
SECOND GEOMETRY 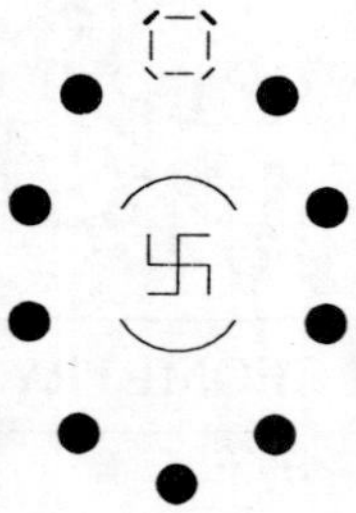	THIRD GEOMETRY 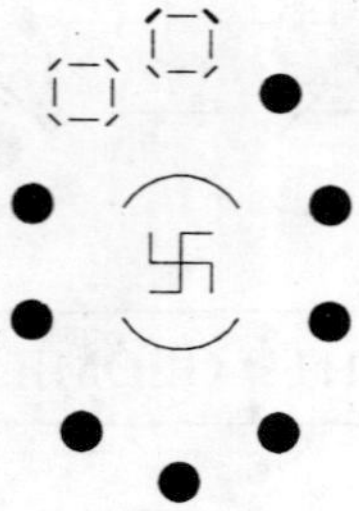
FOURTH GEOMETRY 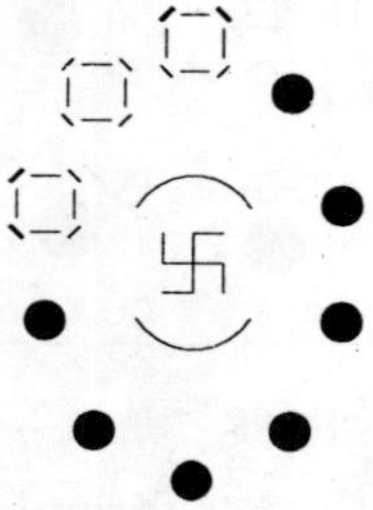	FIFTH GEOMETRY 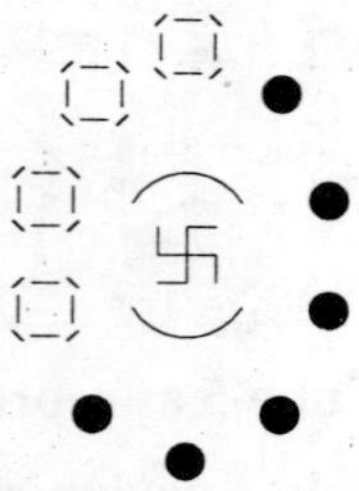

SIXTH GEOMETRY	SEVENTH GEOMETRY
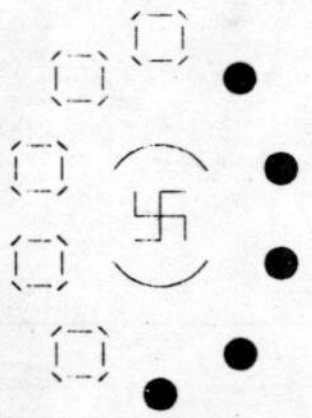	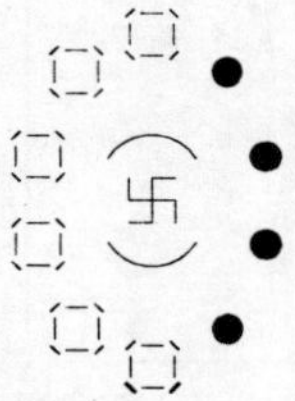
EIGHTH GEOMETRY	NINTH GEOMETRY
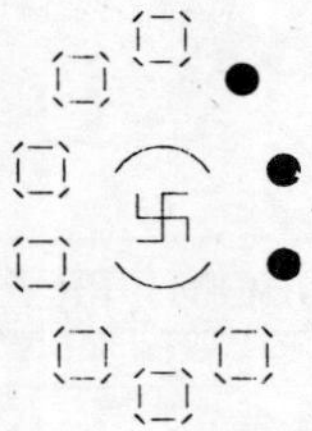	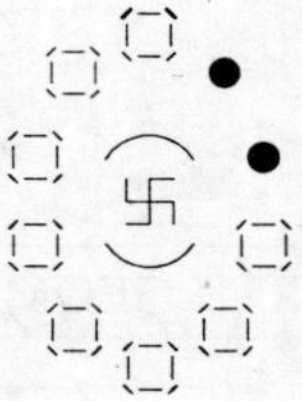
TENTH GEOMETRY	ELEVENTH GEOMETRY
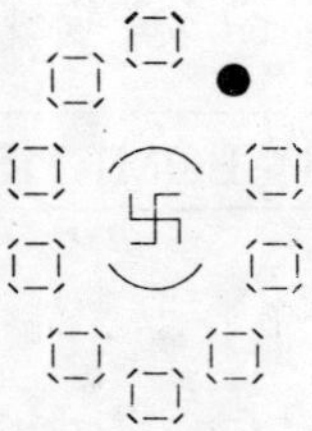	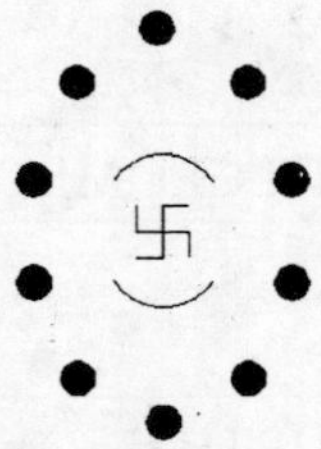

Hyper Cube-6 as representative body of 6-Space

The domain boundary ratio of Hyper cube-6 accepts formulation A^6:$12B^5$

As such Hyper Cube-6 has thirteen versions, namely, Hyper Cube-6 with all the twelve hyper solid-5 boundary components, and with eleven, ten, nine, eight, seven, six, five, four, three, two, one and no hyper solid-5 boundary component.

These thirteen versions of Hyper Cube-6 are representative bodies of 13 geometries of 6-space.

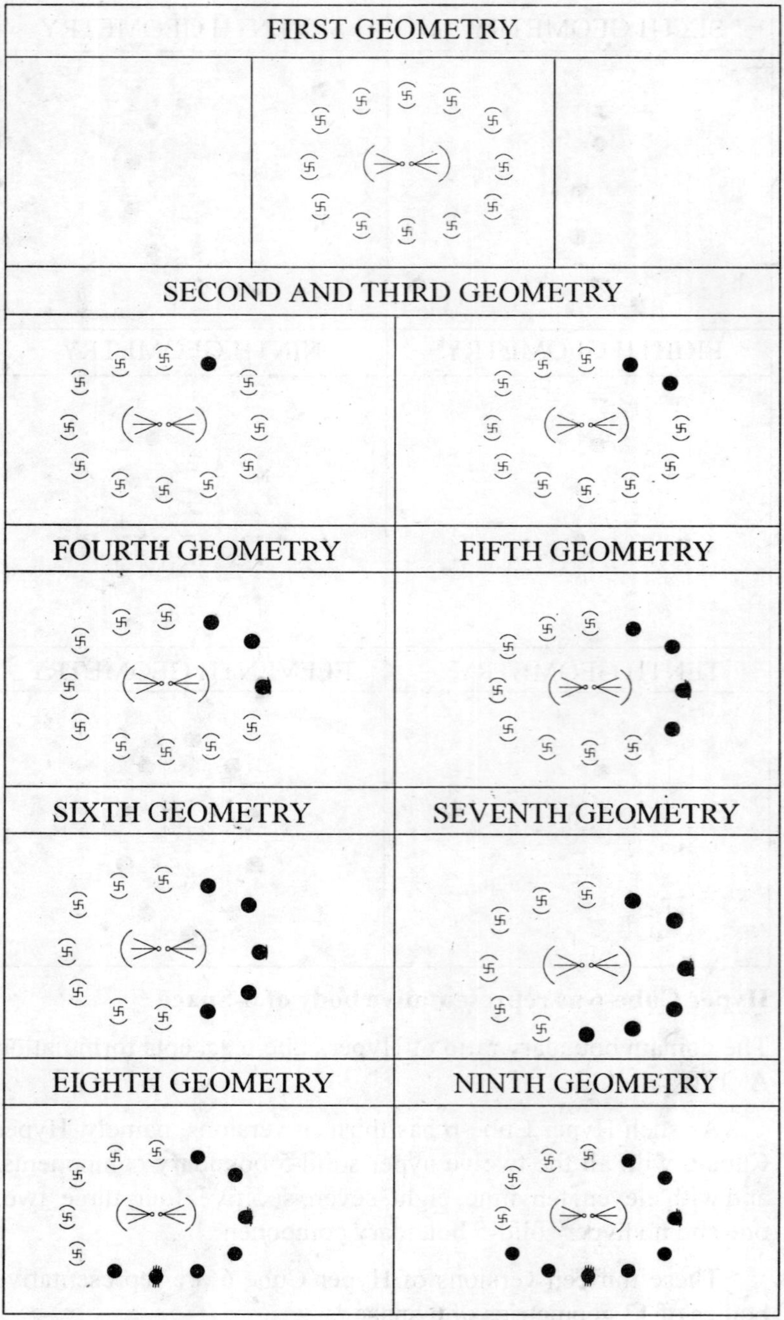

FIRST GEOMETRY

SECOND AND THIRD GEOMETRY

FOURTH GEOMETRY

FIFTH GEOMETRY

SIXTH GEOMETRY

SEVENTH GEOMETRY

EIGHTH GEOMETRY

NINTH GEOMETRY

TENTH GEOMETRY	ELEVENTH GEOMETRY
TWELFTH GEOMETRY	THIRTEENTH GEOMETRY

Hyper Cube-N as representative body of N-Space

The domain boundary ratio of Hyper cube-N accepts formulation A^N:$2NB^{N-1}$

As such Hyper Cube-N has (2N+1) versions, namely, Hyper Cube-N with all the 2N hyper solid-(N-1) boundary components, and with 2N-1, 2N –2, …………….. eleven, ten, nine, eight, seven, six, five, four, three, two, one and no hyper solid-(N-1) boundary component.

These 2N+1 versions of Hyper Cube-N are representative bodies of 2N+1 geometries of N-space.

*****

Lesson 11

Requirement of 960 Cubes to net 6-space Domain

The domain boundary ratio $B^n : 2nA^{n-1}$, sequentially gives us following ratios: $B^6 : 12A^5$, $B^6 : 120A^4$, $B^6 : 960A^4$.

We can invent symbols for hypercubes-6, 5 and 4 in continuity of cube, square and interval. Here with the help of invented symbols for hypercubes 6, 5 and 4, the above requirement of 960 cubes for netting 6-space domain is being depicted:

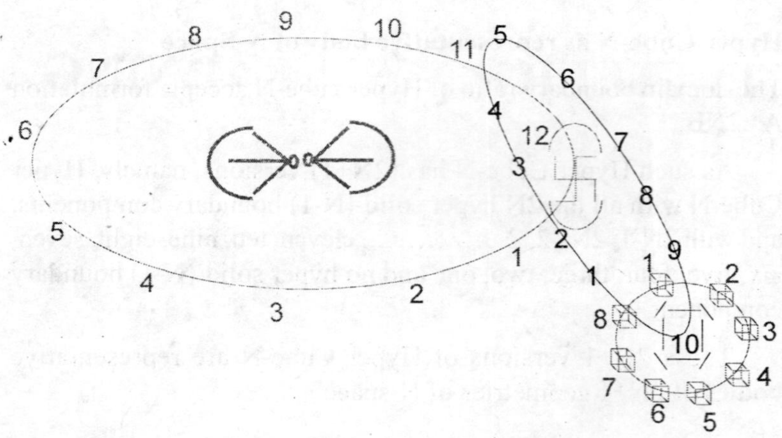

Note:

Physical approach (as physical lump / content / 3-space domain) to anything happening within the solar zone can be netted in terms of 960 characteristics. This geometric format is potent enough to explain the existence phenomenon of human beings as depictable in terms of the human frame / human body.

*****

Section VIII

FOR COSMIC INTELLIGENCE LEARNING FROM STAGE-1

Stage-One

1. Text
2. Mathematics Activity

1
Text

	Text	Devnagri Script	Roman Script
	Ganita Sutra-1	एकाधिकेन पूर्वेण	*Ekadhiken Purvena*

<table>
<tr><td rowspan="2">1. Working rule</td><td>पहले वाले से एक अधिक</td><td>One more than previous one.</td></tr>
<tr><td colspan="2">1, 1 + 1, (1 + 1) + 1, { (1 + 1) + 1} + 1, ...</td></tr>
<tr><td rowspan="2">2. Format</td><td>एकदेशीय-रेखा</td><td>One space-line</td></tr>
<tr><td colspan="2">—●—●—●—●—●——————●——</td></tr>
</table>

2
Mathematics Activity

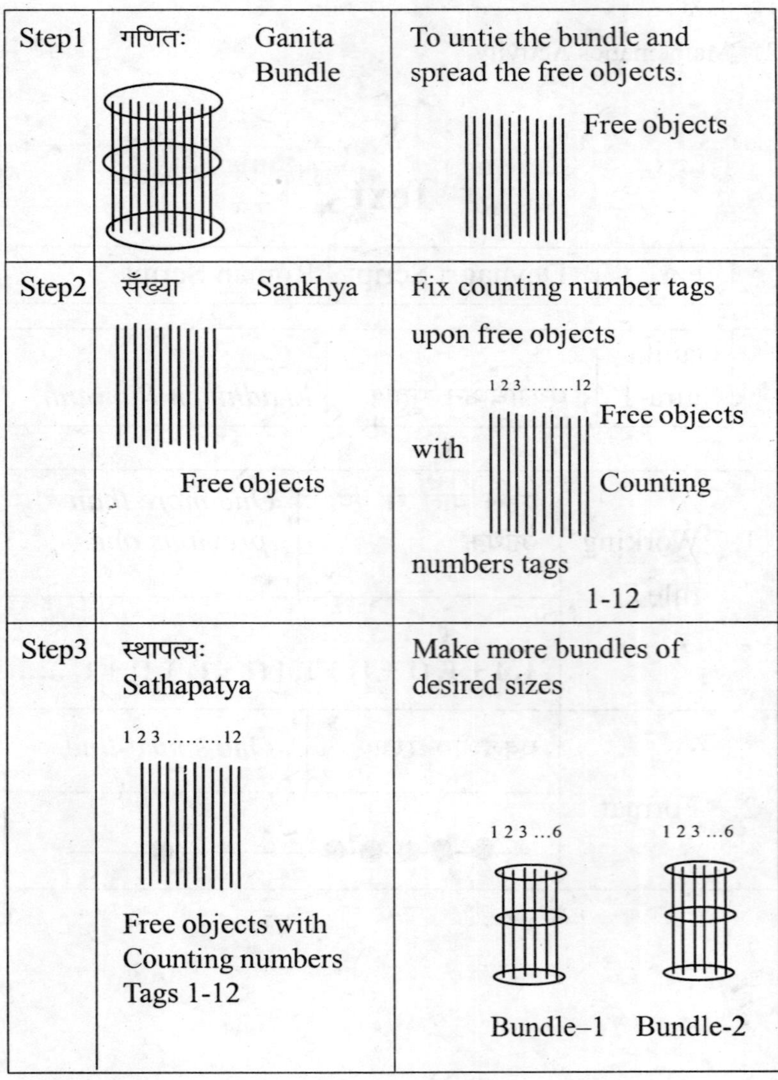

Step1	गणितः Ganita Bundle	To untie the bundle and spread the free objects. Free objects
Step2	सँख्या Sankhya Free objects	Fix counting number tags upon free objects 1 2 3 ....12 Free objects with Counting numbers tags 1-12
Step3	स्थापत्यः Sathapatya 1 2 3 ....12 Free objects with Counting numbers Tags 1-12	Make more bundles of desired sizes 1 2 3 ...6 1 2 3 ...6 Bundle–1 Bundle-2

LEARN VEDIC MATHEMATICS FOR COSMIC INTELLIGENCE

STAGE-ONE

(CLASS-1)

TEXT

	Text	Devnagri Script	Roman Script
	Ganita Sutra-1	एकाधिकेन पूर्वेण	*Ekadhiken Purvena*

1. Working rule	पहले वाले से एक अधिक	*One more than previous one.*
	1, 1 + 1, (1 + 1) + 1, { (1 + 1) + 1} + 1, ...	
2. Format	एकदेशीय–रेखा	*One space-line*
	—●●●●●————●—	

Lesson 1

Counting With Rule from 1 to 10

1.1. Rule is: One more than previous one.

1.2. We know counting uptil ten as one, two, three, four, five, six, seven, eight, nine, ten.

1.3. We can count and write these ten counts as:

One	Two	Three	Four	Five	Six	Seven	Eight	Nine	Ten
1	2	3	4	5	6	7	8	9	10

1.4. Let us start with **one**. We can reach at next count "**two**" with the help of rule "one more than previous one" as:

Step-1	Step-2	Step-3
Fix previous one	Add one under the rule	Reach at next count
One	One+One	Two
1	1+1	2

1.5. Let us start with **two**. We can reach at next count "**three**" with the help of rule "one more than previous one" as:

Step-1	Step-2	Step-3
Fix previous one	Add one under the rule	Reach at next count
Two	Two+One	Three
2	2+1	3

1.6. Let us start with **three**. We can reach at next count "**four**" with the help of rule "one more than previous one" as:

Step-1	Step-2	Step-3
Fix previous one	Add one under the rule	Reach at next count
Three	Three+One	Four
3	3+1	4

1.7. Let us start with **Four**. We can reach at next count "**Five**" with the help of rule "one more than previous one" as:

Step-1	Step-2	Step-3
Fix previous one	Add one under the rule	Reach at next count
Four	Four+One	Five
4	4+1	5

1.8. Let us start with **Five**. We can reach at next count "**Six**" with the help of rule "one more than previous one" as:

Step-1	Step-2	Step-3
Fix previous one	Add one under the rule	Reach at next count
Five	Five+One	Six
5	5+1	6

1.9. Let us start with **Six**. We can reach at next count "**Seven**" with the help of rule "one more than previous one" as:

Step-1	Step-2	Step-3
Fix previous one	Add one under the rule	Reach at next count
Six	Six+One	Seven
6	6+1	7

1.10. Let us start with **Seven**. We can reach at next count "**Eight**" with the help of rule "one more than previous one" as:

Step-1	Step-2	Step-3
Fix previous one	Add one under the rule	Reach at next count
Seven	Seven+One	Eight
7	7+1	8

1.11.Let us start with **Eight**. We can reach at next count "**Nine**" with the help of rule "one more than previous one" as:

Step-1	Step-2	Step-3
Fix previous one	Add one under the rule	Reach at next count
Eight	Eight+One	Nine
8	8+1	9

1.12. Let us start with **Nine**. We can reach at next count "**Ten**" with the help of rule "one more than previous one" as:

Step-1	Step-2	Step-3
Fix previous one	Add one under the rule	Reach at next count
Nine	Nine+One	Ten
9	9+1	10

1.13. Let us start with count **Ten (10)**. We can reach at next count. We can name this count as "**Eleven**" and write this count as 11.

Step-1	Step-2	Step-3
Fix previous one	Add one under the rule	Reach at next count
Three	Three+One	Four
3	3+1	4

1.14. CHASE EXERCISE

Chase on the above lines counting numbers one to ten with the help of the Rule of the Sutra.

*****

Lesson 2
Numbers Line

HAVE ONE MORE PICK

2.1 The Ganita Sutra rule “One more than the previous one” can be applied for counting the pebbles in the basket by picking up the pebbles one by one.

2.2 For this exercise we may have one basket full of pebbles and another empty basket in which the pebbles picked from the full basket are to be put.

2.3 The counting exercise to be performed step wise as under:

STEP-1 Pick One pebble from the first basket and put it in second basket.

STEP-2 Have one more pick of pebble from the first basket and put it in the second basket.

STEP-3 Repeat step two.

STEP-4 Continue repeating step two till the last pebble in the basket.

2.4 These steps of the exercise permit chase in terms of the rule of the Sutra and when the exercise steps are chased under the rule of the Sutra, as a result, the pebbles would get counting numbers tags and in fact would become the counting pebbles.

SETTING OF COUNTING PEBBLES

2.5 Setting of counting pebbles along a line makes the line a **numbers line.**

2.6 Straight pebbles can be picked up from the basket one by one under the rule of one more pick of the pebble till the last pebble and at every pick the picked pebble can be placed one after the other along the line in the sequence of the picks.

2.7 This activity and exercise following the above exercise of shifting pebbles from first basket to another basket and thereby making the pebbles as counting pebbles and then counting pebbles arranged as a setting of counting pebbles in the sequence and order of the counting numbers constructed under the rule of Sutra and thereby construction of numbers line is the biggest achievement for which the students can feel very happy as having achievement.

CHASE EXERCISES

Above both exercises that perform by repeating the same with different number of pebbles in the first basket.

*****

Lesson 3

Counting with Rule From ten to nineteen 10, 11, 12, 13, 14, 15, 16, 17, 18, 19

RE-CAPITULATION

3.1 Let us re-capitulate from learning of lesson-One, 1.13, the way the count following count ten has been reached at as:

3.1.1 Let us start with count **Ten (10)**. We can reach at next count. We can name this count as "**Eleven**" and write this count as 11.

Step-1	Step-2	Step-3
Fix previous one	Add one under the rule	Reach at next count
Ten	Ten+One	Eleven
10	10+1	11

COUNTS AS TWO DIGITS EXPRESSIONS

3.2 Let us see that count ten (10) is expression of two digits.

Count	Second Digit	First Digit
Ten	One	Zero
10	1	0

3.3 Likewise we may observe that the counts following ten namely count eleven is also a two-digit expression.

Count	Second Digit	First Digit
Eleven	One	One
11	1	1

3.4 Now if we simultaneously observe the expressions of counts ten and eleven, we may observe the pattern as that while in first digits there is a shift from zero to one but in second digit there is no change as in both cases same value as of one has been availed.

Count	Second Digit	First Digit
10	1	0
11	1	1

3.5 We may observe that the first digits of counts ten and eleven are themselves obeying the rule of sutra-1. The rule works here in case of first digits as that when the previous count is zero, the following count comes to be zoro+one equal to One.

This chase for first digits of counts ten and eleven can be depicted their stepwise chase as under:

Step-1	Step-2	Step-3
Fix previous one	Add one under the rule	Reach at next count
Zero	Zero+One	One
0	0+1	1

3.6 Now if we proceed further from count eleven to its next count we shall be observing as that:

Step-1	Step-2	Step-3
Fix previous one	Add one under the rule	Reach at next count
Eleven	Eleven+One	Twelve
11	11+1	12

3.7 When the two digits expressions for counts ten, eleven and twelve are observed simultaneously, it would emerge as under:

Count	Second Digit	First Digit
10	1	0
11	1	1
12	1	2

3.8 We may notice that the second digit remain same but in first digit there is increase of one as of rule of Sutra for a shift from count 1 to count 2.

3.9 From the following table we may observe that the first digits of counts ten to nineteen are following the rule of counts zero to nine:

Count	Second Digit	First Digit
10	1	0
11	1	1
12	1	2
13	1	3
14	1	4
15	1	5
16	1	6
17	1	7
18	1	8
19	1	9

CONCLUSION

3.10 Conclusion comes to be that our successful chase of counts Zero to Nine in terms of rule of the sutra, would help us straight a way write two digits expressions for counts ten to nineteen by maintaining second digit at a constant value of count-1 while the first digits would be as and in order of counts zero to nine.

CHASE EXERCISE

3.11 Chase on the above lines counting numbers ten to nineteen with the help of the Rule of the Sutra.

*****

Lesson 4

Counting with Rule From 20 to 29

RECAPITULATION

4.1 We may recapitulate from previous lessons as to how chase of counts 1 to 10 can be availed for chase of counts ten to nineteen by maintaining second digit at constant value of count-1 while the first digits to unfold as counts zero to nine.

Count	Second Digit	First Digit
10	1	0
11	1	1
12	1	2
13	1	3
14	1	4
15	1	5
16	1	6
17	1	7
18	1	8
19	1	9

TWO DIGITS EXPRESSIONS FOR COUNTS ONE TO NINE

4.2 Here below is tabulated two digits expressions for counts One to Nine.

Single digit Count as two digit Expression	Second Digit	First Digit
1 = 01	0	1
2 = 02	0	2
3 = 03	0	3
4 = 04	0	4
5 = 05	0	5

6 = 06	0	6
7 = 07	0	7
8 = 08	0	8
9 = 09	0	9

4.3 Like wise, Zero as a single digit count can be expressed as two digits as under:

Single digit Count as two digit Expression	Second Digit	First Digit
0 = 00	0	0

4.4 We can make Counts Zero to Nine as a bundle of ten counts with distinction as that its all the ten counts accept Zero as a common second digit.

First Bundle of ten counts from Zero to Nine.

Single digit Count as two digit Expression	Second Digit	First Digit
0 = 00	0	0
1 = 01	0	1
2 = 02	0	2
3 = 03	0	3
4 = 04	0	4
5 = 05	0	5
6 = 06	0	6
7 = 07	0	7
8 = 08	0	8
9 = 09	0	9

TWO DIGITS EXPRESSIONS FOR SECOND BUNDLE OF TEN COUNTS

FROM TEN TO NINETEEN (10, 11, 12, 13, 14, 15, 16, 17, 18, 19)

4.5 We can observe that ten counts from ten to nineteen constitute second bundle of counts.

4.6 We can further observe that this second bundle of ten counts (10, 11, 12, 13, 14, 15, 16, 17, 18, 19) has the distinction as that its all the ten counts accept **one** as a common second digit.

4.7 Now if we observe simultaneously both the bundles that is the first bundle of counts 01, 02, 03, 04, 05, 06, 07, 08, 09 and the second bundle of next ten counts 10, 11, 12, 13, 14, 15, 16, 17, 18, 19, it emerges as that while the special feature of first bundle is that it has common value for second digits as **Zero,** on the other hand the special feature of second bundle is that it has common value for digits as **ONE**.

4.8 Conclusion comes to be that the special features of first bundle and second bundle of counts, in terms of common values of their second digits come to be **Zero** and **One** counts respectively.

MAKING TEN BUNDLES OF TEN COUNTS EACH

4.9 The special feature of first bundle of ten counts (00, 01, 02, 03, 04, 05, 06, 07, 08, 09) as Zero value count for their second digits and special feature of second bundle of ten counts (10, 11, 12, 13, 14, 15, 16, 17, 18, 19) as One value count for their second digits can be extended under the rule of Ganita Sutra as special features for ten bundles of ten counts each from 00 to 99 as tabulated here under:

Sr. No.	Bundle	Ten counts	Second digit	Sutra rule for Special feature
1.	First	01, 02, 03, 04, 05, 06, 07, 08, 09	0	Starting Position
2.	Second	10, 11, 12, 13, 14, 15, 16, 17, 18, 19	1	0+1
3.	Third	20, 21, 22, 23, 24, 25, 26, 27, 28, 29	2	1+1
4.	Fourth	30, 31, 32, 33, 34, 35, 36, 37, 38, 39	3	2+1
5.	Fifth	40, 41, 42, 43, 44, 45, 46, 47, 48, 49	4	3+1
6.	Sixth	50, 51, 52, 53, 54, 55, 56, 57, 58, 59	5	4+1
7.	Seventh	60, 61, 62, 63, 64, 65, 66, 67, 68, 69	6	5+1
8.	Eighth	70, 71, 72, 73, 74, 75, 76, 77, 78, 79	7	6+1
9.	Ninth	80, 81, 82, 83, 84, 85, 86, 87, 88, 89	8	7+1
10	Tenth	90, 91, 92, 93, 94, 95, 96, 97, 98, 99	9	8+1

ACHIEVEMENT

4.10 Students can feel very happy with their achievements of Mathematical chase of hundred counts constructed with Ganita Sutra rule.

4.11 Two digits expressions as ten rows and ten columns is double achievements, firstly constructions of counts with Mathematical rule and secondly two digits expressions also standing chased Mathematically.

4.12 Further achievement is to have a format of ten rows and ten columns.

4.13 This achievement of format of rows and columns is wonderful achievement over the linear expressions for the count along the line.

CHASE EXERCISE

4.14 Step by step chase the construction of hundred counts for their expressions in two digits form on ten rows and ten columns format.

*****

VEDIC MATHEMATICS

(For Cosmic Intelligence) Outline of Four Weeks Training Course for First Stage Vedic Mathematics Teachers

First Stage:
Outline Contents

1. **GENERAL OUTLINE OF FOUR WEEKS TRAINING COURSE OF VEDIC MATHEMATICS TEACHERS**

2. **GENERAL INTRODUCTION OF THE COURSE**

    1. Ganita Sutras focus
    2. Vedic Geometric focus
    3. Text of Ganita Sutra-1
    4. Formal steps of Mathematics Education
    5. Potentialities of the child to be availed for teaching
    6. Mathematics Content
    7. Mathematical tool
    8. Applied value
    9. Basis Basics of learning and teaching
    10 Skills
    11. Applied skills

3. **LESSONS**

    GROUP-1 LESSONS (1 To 10)
    (FIRST SEMESTER FIRST STAGE FIRST WEEK
    Vedic Mathematics Teachers Training Course)

Lesson-1 Oral counting.

Lesson-2 Symbols writing appraisal.

Lesson-3 Matching of counts one to ten with their writing symbols.

Lesson-4 Setting of counting pebbles in slots of numbers line.

Lesson-5 Repeated picking up of 2 pebbles at a time and their settings.

Lesson-6 Repeated picking up of 3 pebbles at a time and their setting.

Lesson-7 Repeated picking up of 4 pebbles at a time and their setting.

Lesson-8 Repeated picking up of 5 and 6 pebbles as alternative picks and their setting.

Lesson-9 Repeated picking up of 7 and 8 pebbles as alternative picks and their setting.

Lesson-10 Repeated picking up of 9 and 10 pebbles as alternative picks and their setting.

GROUP-I1 LESSONS (11 To 20)
(FIRST SEMESTER FIRST STAGE SECOND WEEK
Vedic Mathematics Teachers Training Course)

Learn & Teach

VEDIC MATHEMATICS

TEXT BOOK
(For Cosmic Intelligence)

CLASS-1

FOUR WEEKS TRAINING PROGRAMME

FIRST WEEK

First day

1. Introduction to the training programme and training stage.
2. Ideal expectation of the total programme and of the present's stage during its first week.
3. Overview of learning and teaching during this week.
4. Vedic way to approach the knowledge.
5. The way Intellectual approach and transcendental approach to go together.
6. Beginning of formal Mathematics education.
7. "Innocent intellect handling responsibility."
8. Aim is to shape the intelligence.
9. Goal is to attain cosmic order.
10. Self-appraisal exercises at the end of each stage of learning.

Second day

1. Mathematics range and time frame.
2. Four weeks learning to match forty weeks teaching.
3. This week learning to match first ten weeks teaching.
4. First ten weeks teaching course of class-I.

5. Learning 1st week teaching course of class-I: Lesson-1.
6. Learning 2nd week teaching course of class-I: Lesson-2

Third day

1. Learning 3rd week teaching course of class-I: Lesson-3
2. Learning 4th week teaching course of class-I: Lesson-4
3. Learning 5th week teaching course of class-I: Lesson-5
4. Learning 6th week teaching course of class-I: Lesson-6

Fourth day

1. Learning 7th week teaching course of class-I: Lesson-7
2. Learning 8th week teaching course of class-I: Lesson-8
3. Learning 9th week teaching course of class-I: Lesson-9
4. Learning 10th week teaching course of class-I: Lesson-10

Fifth day

Trainees appraisals interaction and the lessons of the training of the week.

SECOND WEEK

First day

1. Introduction to the training stage.
2. Expectation of the present's stage during this week.
3. Overview of learning and teaching during this week.
4. Background of learning of the previous week.
5. Proceeding in continuity of the previous week.
6. Continuous exercise of shaping of intelligence.
7. Goal of learning during this week.
8. Projection of the shape of situations to be met during this week of learning range.

Second day

1. Mathematics range and time frame.
2. Four weeks learning to match forty weeks teaching.
3. This week learning to match first ten weeks teaching.
4. First ten weeks teaching course of class-I.
5. Learning 11th week teaching course of class-I: Lesson-11
6. Learning 12th week teaching course of class-I: Lesson-12

Third day

1. Learning 13th week teaching course of class-I: Lesson-13
2. Learning 14th week teaching course of class-I: Lesson-14
3. Learning 15th week teaching course of class-I: Lesson-15
4. Learning 16th week teaching course of class-I: Lesson-16

Fourth day

1. Learning 17th week teaching course of class-I: Lesson-17
2. Learning 18th week teaching course of class-I: Lesson-18
3. Learning 19th week teaching course of class-I: Lesson-19
4. Learning 20th week teaching course of class-I: Lesson-20

Fifth day

Trainees appraisals interaction and the lessons of the training of the week.

THIRD WEEK

First day

1. Introduction to the training stage.
2. Expectation of the present's stage during this week.
3. Overview of learning and teaching during this week.

4. Background of learning of the previous weeks.
5. Proceeding in continuity of the previous weeks.
6. Continuous exercise of shaping of intelligence.
7. Goal of learning during this week.
8. Projection of the shape of situations to be met during this week of learning range.

Second day

1. Mathematics range and time frame.
2. Four weeks learning to match forty weeks teaching.
3. This week learning to match first ten weeks teaching.
4. First ten weeks teaching course of class-I.
5. Learning 21st week teaching course of class-I: Lesson-21
6. Learning 22nd week teaching course of class-I: Lesson-22

Third day

1. Learning 23rd week teaching course of class-I: Lesson-23
2. Learning 24th week teaching course of class-I: Lesson-24
3. Learning 25th week teaching course of class-I: Lesson-25
4. Learning 26th week teaching course of class-I: Lesson-26

Fourth day

1. Learning 27th week teaching course of class-I: Lesson-27
2. Learning 28th week teaching course of class-I: Lesson-28
3. Learning 29th week teaching course of class-I: Lesson-29
4. Learning 30th week teaching course of class-I: Lesson-30

Fifth day

Trainees appraisals interaction and the lessons of the training of the week.

FOURTH WEEK

First day

1. Introduction to the training stage.
2. Expectation of the present's stage during this week.
3. Overview of learning and teaching during this week.
4. Background of learning of the previous weeks.
5. Proceeding in continuity of the previous weeks.
6. Continuous exercise of shaping of intelligence.
7. Goal of learning during this week.
8. Projection of the shape of situations to be met during this week of learning range.

Second day

1. Mathematics range and time frame.
2. Four weeks learning to match forty weeks teaching.
3. This week learning to match first ten weeks teaching.
4. First ten weeks teaching course of class-I.
5. Learning 31st week teaching course of class-I: Lesson-31
6. Learning 32nd week teaching course of class-I: Lesson-32
7. General appraisal of this day's learning.

Third day

0. General appraisal of previous day's learning.
1. Learning 33rd week teaching course of class-I: Lesson-33
2. Learning 34th week teaching course of class-I: Lesson-34
3. Learning 35th week teaching course of class-I: Lesson-35
4. Learning 36th week teaching course of class-I: Lesson-36
5. General appraisal of this day's learning.

Fourth day

0. General appraisal of previous day's learning.
1. Learning 37th week teaching course of class-I: Lesson-37
2. Learning 38th week teaching course of class-I: Lesson-38
3. Learning 39th week teaching course of class-I: Lesson-39
4. Learning 40th week teaching course of class-I: Lesson-40
5. General appraisal of this day's learning.

Fifth day General appraisal of the first stage training course.

*****

Learn & Teach
VEDIC MATHEMATICS
TEXT BOOK
(For Cosmic Intelligence)

CLASS-1

1. GANITA SUTRAS FOCUS

Ganita Sutra	Space
Ganita sutra 1	1-Space

2. VEDIC GEOMETRIC FOCUS

	1-Space
…..	___
	Line

3. TEXT OF GANITA SUTRA 1

Devnagri Script	एकाधिकेन पूर्वेण
Roman Script	Ekadhikena Purven
English Rendering	One more than the One Before

4. FORMAL STEPS OF MATHEMATICS EDUCATION

Here formal Vedic Mathematics education of child begins He is formally initiated for shaping of his intelligence in terms of the discipline and order of Ganita-Sutra-1. The Mathematics domain and content planned and designed as formal course of instructions is the application skill of working rule of Ganita Sutra-1 for construction of counting numbers and their setting in that sequence and order along a line (designated as counting numbers line).

5. POTENTIALITIES OF THE CHILD TO BE AVAILED FOR TEACHING

1. Child has innocent memory, which admits sound signals for its desirable shaping and retention for communication as sound signals through self-speech of the child.

2. Child's intellect shaping and memory retention reflects in communicative system of child which express in display of coordination of organs of his body with command motion of the hand and fingers thereof as the index output, therefore the oral communications with the child can be formalized for formal Mathematics education in channelising the index output by putting it in the stream of opting one out of two symbols making that is one out of tick (✓) and cross (x). Then activity of sorting dissimilar for similars, patterns recognitions, symmetries chase etc. etc. may well become proper communication channels of the teachers with the intellect of the young brains.

6. MATHEMATICS CONTENT

1. Comprehension of concept of ONE
2. Comprehension of concept of ORDER
   in terms of BEFORE AND AFTER.
3. Comprehend the concept of ONE MORE.
4. Comprehend the Mathematics content of working rule
   "BY ONE MORE THAN ONE BEFORE"

7. MATHEMATICAL TOOLS

1. Counting numbers
2. One to one correspondence
3. Addition operation

8. APPLIED VALUE

1. Construction of counting pebbles.

2. Construction of counting numbers line.
3. Construction of sequences of the order of counting numbers.

9. BASIS BASICS OF LEARNING AND TEACHING

1. Confident oral counting from one to hundred.
2. Clear symbols reorganizations for one to ten.
3. Proper writing skill for symbols 0, 1, 2, 3, 4, 5, 6, 7, 8, 9 and 10
4. Sorting with confidence of similar objects from most dissimilar objects like balls from bats.
5. Setting of counting pebbles in slots along a line in counting sequence.
6. Counting of slots of a counting line.
7. One to One correspondence of counting pebbles and counting slots of counting line
8. One to One correspondence between pebbles of two baskets.
9. One to One correspondence of sticks of two bundles.
10. Equal sharing of fruits by two friends.
11. Equal sharing of fruits between three friends.
12. Comparison by counting of sticks of two bundles.
13. Comparison by counting of pebbles of two baskets.

10. SKILLS

1. Counting from one to hundred.
2. Setting of counting pebbles along a line.
3. Picking up two pebbles at a time.
4. Picking up three pebbles at a time.
5. Picking up four pebbles at a time.

6. Picking up five pebbles at a time.
7. Untying bundle of ten sticks and counting the sticks. And making them,
   (a) two bundles of five sticks each.
   (b) five bundles of two sticks each.
8. Untying bundle of twenty sticks and counting the sticks. And making them,
   (a) two bundles of ten sticks each.
   (b) four bundles of five sticks each.
   (c) Five bundles of four sticks each.
   (d) Ten bundles of two sticks each.
9. Untying bundle of thirty sticks and counting the sticks. And making them,
   (a) two bundles of fifteen sticks each.
   (b) three bundles of ten sticks each.
   (c) five bundles of six sticks each.
   (d) six bundles of five sticks each.
   (e) ten bundles of three sticks each.
10. Untying bundle of Forty sticks and counting the sticks. And making them,
   (a) two bundles of twenty sticks each.
   (b) four bundles of ten sticks each.
   (c) five bundles of eight sticks each.
   (d) four bundles of ten sticks each.
   (e) eight bundles of five sticks each.
   (f) ten bundles of four sticks each.
11. Untying bundle of fifty sticks and counting the sticks. And making them,
   (a) two bundles of twenty-five sticks each.

(b) five bundles of ten sticks each.

(c) ten bundles of five sticks each.

12. Untying bundle of sixty sticks and counting the sticks. And making them,

    (a) two bundles of thirty sticks each.

    (b) three bundles of twenty sticks each.

    (c) five bundles of twelve sticks each.

    (d) ten bundles of six sticks each.

    (e) twelve bundles of five sticks each.

    (f) fifteen bundles of four sticks each.

    (g) thirty bundles of two sticks each.

13. Untying bundle of seventy sticks and counting the sticks. And making them,

    (a) two bundles of thirty-five sticks each.

    (b) five bundles of fourteen sticks each.

    (c) seven bundles of seven sticks each.

    (d) ten bundles of seven sticks each.

    (e) fourteen bundles of five sticks each.

    (f) thirty-five bundles of two sticks each.

14 Untying bundle of eighty sticks and counting the sticks. And making them,

    (a) two bundles of forty sticks each.

    (b) four bundles of twenty sticks each.

    (c) five bundles of sixteen sticks each.

    (d) eight bundles of ten sticks each.

    (e) ten bundles of eight sticks each.

    (f) sixteen bundles of five sticks each.

    (g) twenty bundles of four sticks each.

15 Untying bundle of ninety sticks and counting the sticks. And making them,

(a) two bundles of forty-five sticks each.

(b) three bundles of thirty sticks each.

(c) five bundles of eighteen sticks each.

(d) nine bundles of ten sticks each.

(e) ten bundles of nine sticks each.

(f) eighteen bundles of five sticks each.

(g) thirty bundles of three sticks each.

(h) forty-five bundles of two sticks each.

16 Untying bundle of hundred sticks and counting the sticks. And making them,

(a) two bundles of fifty sticks each.

(b) five bundles of twenty sticks each.

(c) ten bundles of ten sticks each.

(d) twenty bundles of five sticks each.

(e) fifty bundles of two sticks each.

11. APPLIED SKILLS

1. Money	To distinguish the coins and currency.
2. Time	To locate position numbers of big and small hand of the clock.
3. Dinning table	To serve each member one item of fruit.
4. Drawing room	Count the number of chairs and guests in the house.
5. Birthday party	Carry number of toffees as is the number of the classmates.
6. Day of week	Count with the help of bundle of sevens.

*****

Learn & Teach
VEDIC MATHEMATICS
TEXT BOOK
(For Cosmic Intelligence)

CLASS-1 FORTY LESSONS COURSE

GROUP ONE: LESSON 1-10

(To be covered during first week of first semester of the Vedic Mathematics teacher training course)

LESSON-1 ORAL COUNTING

First day

1. Oral counting appraisal of previous learning
2. Oral introduction of rule of Ganita Sutra-1

Second day — Counting uptil 50

1. Step by step with rule of Sutra-1
2. With different objects of different settings

Third day — Counting uptil 70

1. Step by step with rule of Sutra-1
2. With different objects of different settings

Fourth day — Counting uptil 100

1. Step by step with rule of Sutra-1
2. With different objects of different settings

Fifth day	Appraisal of learning during this week with focus upon the comprehension of step by step increase of counts as per rule of Sutra-1.

LESSON-2 SYMBOLS WRITING APPRAISAL

First day	Appraisal of previous learning 1. Of symbols 1 to 9 and 0. 2. Recitation of Sutra-1.
Second day	Writing of 10 to 19 (Step by step writing with oral recitation of rule of Sutra-1 taking from 10 to 19)
Third day	Writing of 20 to 29 (Step by step writing with oral recitation of rule of Sutra-1 taking from 20 to 29)
Fourth day	Writing of two symbols together. (Step by step writing from any count onwards with oral recitation of rule of Sutra-1)
Fifth day	Appraisal of learning during this week with focus upon the parallel comprehension of writing skill and step by step increase of counts as per rule of Sutra-1.

LESSON-3 MATCHING OF COUNTS ONE TO TEN WITH THEIR WRITING SYMBOLS

First day	Appraisal of intuitive instincts of matching of two objects
Second day	Matching of counts One to Ten and their symbols 1 to 10

Third day	Matching of two bundles of ten objects with the rule of Ganita Sutra-1, first matching one object of one bundle with one object of another bundle, then taking one more object from the first bundle and matching it with one more object of the second bundle and so on.
Fourth day	Matching of ten counting pebbles of one basket with ten counting pebbles of another basket. with the rule of Ganita Sutra-1, first matching one pebble of one basket with one pebble of another basket, then taking one more pebble from the first basket and matching it with one more object of the second basket and so on.
Fifth day	Appraisal of learning during this week with focus upon the parallel comprehension of counting numbers and counting numbers line and slots upon the numbers line to be worked out step by step increase of counts as per rule of Sutra-1.

LESSON-4 SETTING OF COUNTING PEBBLES IN SLOTS OF NUMBERS LINE

First day

1. Appraisal of intuitive instincts of setting objects with the help of the rule of Ganita Sutra-1.
2. Construction of number line with the help of rule of Ganita Sutra-1.

Second day	Setting of ten counting pebbles in slots of number line with the help of Ganita Sutra-1.
Third day	Setting of twenty counting numbers in slots of number line with the help of Ganita Sutra-1.
Fourth day	Setting of ten counting pebbles in alternative slots of number line with the help of Ganita Sutra-1.
Fifth day	Appraisal of learning during this week with focus upon the parallel comprehension of counting numbers and counting numbers line and slots upon the numbers line to be worked out step by step increase of counts as per rule of Sutra-1.

LESSON-5 REPEATED PICKING UP OF TWO PEBBLES AT A TIME AND THEIR SETTINGS

First day	1. Appraisal of intuitive instincts of repeated picks. 2. As format of Ganita sutra-1 and Upsutra-1.
Second day	Repeated picks of two pebbles from a basket as formats of Ganitasutra-1 and Upsutra-1.
Third day	Repeated picks of two objects from open courtyard as formats of Ganitasutra-1 and Upsutra-1.
Fourth day	Setting of repeated picks of two counting pebbles in slots of number line as formats of Ganita sutra-1 and Upsutra-1.

Fifth day	Appraisal of learning during this week as formats of Ganita sutra-1 and Up sutra-1.

LESSON-6 REPEATED PICKING UP OF THREE PEBBLES AT A TIME AND THEIR SETTING

First day	Appraisal of intuitive instincts of repeated picks as format of Ganita sutra-1 and Upsutra-1.
Second day	Repeated picks of three pebbles from a basket as format of Ganita sutra-1 and Upsutra-1.
Third day	Repeated picks of three objects from open courtyard as format of Ganita sutra-1 and Upsutra-1.
Fourth day	Setting of repeated picks of three counting pebbles in slots of number line as format of Ganita sutra-1 and Upsutra-1.
Fifth day	Appraisal of learning during this week as format of Ganita sutra-1 and Upsutra-1.

LESSON-7 REPEATED PICKING UP OF FOUR PEBBLES AT A TIME AND THEIR SETTING

First day	Appraisal of intuitive instincts of repeated picks as format of Ganita sutra-1 and Upsutra-1.
Second day	Repeated picks of four pebbles from a basket as format of Ganita sutra-1 and Upsutra-1.
Third day	Repeated picks of four objects from open courtyard as format of Ganita sutra-1 and Upsutra-1.

Fourth day	Setting of repeated picks of four counting pebbles in slots of number line as format of Ganita sutra-1 and Upsutra-1.
Fifth day	Appraisal of learning during this week as format of Ganita sutra-1 and Upsutra-1.

LESSON-8 REPEATED PICKING UP OF FIVE AND SIX PEBBLES AS ALTERNATIVE PICKS AND THEIR SETTING

First day	Appraisal of intuitive instincts of repeated picks as format of Ganita sutra-1 and Upsutra-1.
Second day	Repeated picks of five/six pebbles as alternative picks from a basket as format of Ganita sutra-1 and Upsutra-1.
Third day	Repeated picks of five/six pebbles as alternative picks from open courtyard as format of Ganita sutra-1 and Upsutra-1.
Fourth day	Setting of repeated picks of five/six pebbles as alternative picks counting pebbles in slots of number line as format of Ganita sutra-1 and Upsutra-1.
Fifth day	Appraisal of learning during this week as format of Ganita sutra-1 and Upsutra-1.

LESSON-9 REPEATED PICKING UP OF SEVEN AND EIGHT PEBBLES AS ALTERNATIVE PICKS AND THEIR SETTING

First day	Appraisal of intuitive instincts of repeated picks as format of Ganita sutra-1 and Upsutra-1.
Second day	Repeated picks of seven/eight pebbles as alternative picks from a basket as format of Ganita sutra-1 and Upsutra-1.
Third day	Repeated picks of seven/eight pebbles as alternative picks from open courtyard as format of Ganita sutra-1 and Upsutra-1.
Fourth day	Setting of repeated picks of seven/eight pebbles as alternative picks counting pebbles in slots of number line as format of Ganita sutra-1 and Upsutra-1.
Fifth day	Appraisal of learning during this week as format of Ganita sutra-1 and Upsutra-1.

LESSON-10 REPEATED PICKING UP OF NINE AND TEN PEBBLES AS ALTERNATIVE PICKS AND THEIR SETTING

First day	Appraisal of intuitive instincts of repeated picks as format of Ganita sutra-1 and Upsutra-1.
Second day	Repeated picks of nine/ten pebbles as alternative picks from a basket as format of Ganita sutra-1 and Upsutra-1.
Third day	Repeated picks of nine/ten pebbles as alternative picks from open courtyard as format of Ganita sutra-1 and Upsutra-1.
Fourth day	Setting of repeated picks of nine/ten pebbles as alternative picks

	counting pebbles in slots of number line as format of Ganita sutra-1 and Upsutra-1.
Fifth day	Appraisal of learning during this week as format of Ganita sutra-1 and Upsutra-1.

*****

APPENDIX

Appendix-I

Here below is the list of 126 Lessons of free VM Course No. 1 uploaded on site (www.learn-and-teach-vedic-mathematics.com):

Lessons – Vedic Mathematics on Geometric Formats of Real Spaces

No	Title of the Lesson
1.	Float "SPACE thought" for the TRANSCENDING mind
2.	Chase TRACK of MOVING bodies
3.	Square and Cube and Circle and Sphere
4.	Hyper Cube - 45Synthetic Monad of two Parts
6.	Synthetic Monad of Two Parts
7.	Manifestations Layers
8.	Transcendence Through Manifestations
9.	A Chase of Transcending Mind Through Origin as Spatial Seal
10.	Transcendence Acquiring Fifth Fold
11.	Ascendance with Ambrosia of Bliss of Transcendental World
12.	Vedic Mathematics
13.	Arithmetic to Astronomy
14.	Conscious Attention Makes All the Difference
15.	In-Flow from Surya (Sun) into Akash (Space)
16.	Onward Journey from Akash to Surya
17.	Reaching up till Transcendental Boundary of The Sun
18.	Expanding Universe
19.	Sapat Rishi Lok (7-Space)
20.	Ativahkas of Sapat Rishi Lok Carry to Brahman Lok
21.	Surya To Brahman Domain (6, 7, 8, 9)
22.	Introduction to Vedic Mathematics Science & Technology

23. Three Types of Creations Within Rays of The Sun
24. Pair of Hemispheres 25 Bindu Sarovar
26. (Nad Bindu) Point Reservoir of Sound
27. Tej Bindu Point Reservoir of Light
28. Dhyan Bindu Focus Point of Attention
29. Richness of the format of manifestation layer (3, 4, 5, 6)
30. Different approaches with focus upon 3-Space as dimension
31. Different approaches with focus upon 4-Space as domain
32. Different approaches with focus upon 5-Space as domain
33. Different approaches with 6-Space as origin
34. Different approaches with 7-Space as transcendental base
35. Transcendence through (3, 4, 5, 6) to (4, 5, 6, 7)
36. Om, Parnavah Aum and Onkar
37. Parnavah folds 38 First fold of Parnavah
39. Second fold of Parnavah
40. Third fold of Parnavah
41. Fourth Fold of Parnavah
42. Parnavah form and Formulation
43. Samved Samhita
44. Samved: Purvarchak-Chapter-1
45. Samved: Purvarchak-Chapter-2
46. Samved: Purvarchak-Chapter-3
47. Samved: Purvarchak-Chapter-4
48. Samved: Purvarchak-Chapter-5
49. Samved: Purvarchak-Chapter-6
50. Transcendence into inner folds of transcendental world
51. Transcendence into Dimension of Dimension
52. Transcendence as Trishapta Rule
53. Lord Ganesha, The Lord of Transcendence to Inner Folds
54. Transcendental Phenomena of Transcending and Ascending Mind

55. Spatial Order of Transcendence Format Availed by Transcending and Ascending Mind
56. Mathematics, Science and Technology of Transcendental Phenomena
57. Formats Availed by Vedic Scriptures-1
58. Formats Availed by Vedic Scriptures-2
59. Formats Availed by Vedic Scriptures-3
60. Formats Availed by Vedic Scriptures-4
61. Formats Availed by Vedic Scriptures-5
62. Formats Availed by Vedic Scriptures-6
63. Formats Availed by Vedic Scriptures-7
64. Vedic Systems-1
65. Vedic Systems-2
66. Vedic Systems-3
67. Vedic Systems-4
68. Vedic Systems-5
69. Vedic Systems-6
70. Vedic Systems-7
71. Approach focus: Shift from old formats to new formats-1
72. Approach focus: Shift from old formats to new formats-2
73. Approach focus: Shift from old formats to new formats-3
74. Approach focus: Shift from old formats to new formats-4
75. Approach focus: Shift from old formats to new formats-5
76. Approach focus: Shift from old formats to new formats-6
77. Approach focus: Shift from old formats to new formats-7
78. Approach focus: Reference frame and Absolute frame-1
79. Approach focus: Reference frame and Absolute frame-2
80. Approach focus: Reference frame and Absolute frame-3
81. Approach focus: Reference frame and Absolute frame-4
82. Approach focus: Reference frame and Absolute frame-5
83. Approach focus: Reference frame and Absolute frame-6

84. Approach focus: Reference frame and Absolute frame-7
85. Approach focus: Reference frame and Absolute frame-1
86. Approach focus: Reference frame and Absolute frame-2
87. Approach focus: Reference frame and Absolute frame-3
88. Approach focus: Reference frame and Absolute frame-4
89. Approach focus: Reference frame and Absolute frame-5
90. Approach focus: Reference frame and Absolute frame-6
91. Approach focus: Reference frame and Absolute frame-7
92. Approach focus: First element/Prithvi Tatav/Earth element-Aspect-1
93. Approach focus: First element/Prithvi Tatav/Earth element-Aspect-2
94. Approach focus: First element/Prithvi Tatav/Earth element-Aspect-3
95. Approach focus: First element/Prithvi Tatav/Earth element-Aspect-4
96. Approach focus: First element/Prithvi Tatav/Earth element-Aspect-5
97. Approach focus: First element/Prithvi Tatav/Earth element-Aspect-6
98. Approach focus: First element/Prithvi Tatav/Earth element-Aspect-7
99. Focus: Organization of Vedic Knowledge-1 Initial Information-1
100. Focus: Organization of Vedic Knowledge-1 Initial Information-2
101. Focus: Organization of Vedic Knowledge-1 Initial Information-3
102. Focus: Organization of Vedic Knowledge-1 Initial Information-4
103. Focus: Organization of Vedic Knowledge-1 Initial Information-5

104. Focus: Organization of Vedic Knowledge-1 Initial Information-6
105. Organization of Vedic Knowledge-2 Correlation features 1 and 2
106. Organization of Vedic Knowledge-3 Correlation features 3 and 4
107. Focus: Organization of Vedic Knowledge-2 Correlation features 5 and 6
108. Focus : Organization of Vedic Knowledge-2 Correlation features 7 and 8
109. Focus : Organization of Vedic Knowledge-2 Correlation features 9 and 10
110. Focus : Organization of Vedic Knowledge-2 Correlation features 11
111. Focus: Organization of Vedic Knowledge-2 VMO-1
112. Focus: Organization of Vedic Knowledge-2 VMO-2
113. Focus: Organization of Vedic Knowledge-2 VMO-3
114. Focus: Organization of Vedic Knowledge-2 VMO-4 & 5
115. Focus: Organization of Vedic Knowledge-2 VMO-6 & 7
116. Focus: Organization of Vedic Knowledge-2 VMO-8
117. Focus: Organization of Vedic Knowledge-2 VMO-9
118. Focus: Organization of Vedic Knowledge-2 VMO-10
119. Focus: Organization of Vedic Knowledge-2 VMO-11
120. Focus: Organization of Vedic Knowledge-2 VMO-12
121. Focus: Organization of Vedic Knowledge-2 VMO-13
122. Focus: Organization of Vedic Knowledge-2 VMO-14 & 15
123. Focus: Organization of Vedic Knowledge-2 VMO-16 & 17
124. Focus: Organization of Vedic Knowledge-2 VMO-18 & 19
125. Focus: Organization of Vedic Knowledge-2 VMO-21 & 21
126. Focus: Organization of Vedic Knowledge-2 VMO-22 & 23

*****

Appendix-II

Here below is the list of 112 Lessons of free VM Course No. 2 uploaded on site (www.learn-and-teach-vedic-mathematics.com).

No	Title of the Aspect
1.	About Construction of Ekadhikena Numerals
2.	About Construction of Ekanunena Numerals
3.	About Construction of Negative Numerals
4.	About Construction of Dashadhika Numerals
5.	About Construction of Dashadhika Negative Numerals
6.	Text of Ganita Sutras
7.	Text of Ganita Upsutras
8.	Numerals on A Pair of Digits Format
9.	Arithmetic Operations of Two Digit Numbers: Addition
10.	Skill for multiplication of numerals expressed as pair of digits on geometric format of pair of slots
11.	Skill for multiplication of numbers of two digits on geometric format of a pair of slots
12.	Extension of Format of Pair of Slots into That of Triple Slots
13.	Tables of numerals one, two, three, four and five
14.	Tables of numerals six, seven, eight and nine
15.	Geometric format with mirror at the middle
16.	Reading first book of nature (human body) with the help of a mirror
17.	Organization of two digits number as reflection pairs
18.	Multiplication of 111 x 111
19.	Ganita Upsutra: Anrupena
20.	Ganita Upsutra Viloknam

21. Two digit primes
22. Programme Outline: Class-1
23. Programme Outline: Class-1 Lesson-1
24. Programme Outline: Class-1 Lesson 2
25. Programme Outline: Class-1 Lesson-3
26. Programme Outline: Class-2
27. Programme Outline: Class-3
28. Programme Outline: Class-4
29. Knowledge is a single discipline-1
30. Knowledge is a single discipline-2
31. Knowledge is a single discipline-3
32. Knowledge is a single discipline-4
33. Knowledge is a single discipline-5
34. Knowledge is a single discipline-6
35. Knowledge is a single discipline-7
36. Initiation of Young Minds
37. Mansara
38. Counts and Units
39. Lines and Circles
40. Cube
41. Bodies in Motion
42. Creator's Space
43. Geometry Class in Yaghya Shala-1
44. Geometry Class in Yaghya Shala-2
45. Geometry Class in Yaghya Shala-3
46. Geometry Class in Yaghya Shala-4
47. Geometry Class in Yaghya Shala-5
48. Geometry Class in Yaghya Shala-6
49. Geometry Class in Yaghya Shala-7
50. Ved Shala: Vaidya Shala to Vaiydya shala-1

51. Ved Shala: Vaidya Shala to Vaiydya shala-2
52. Ved Shala: Vaidya Shala to Vaiydya shala-3
53. Ved Shala: Vaidya Shala to Vaiydya shala-4
54. Ved Shala: Vaidya Shala to Vaiydya shala-5
55. Ved Shala: Vaidya Shala to Vaiydya shala-6
56. Ved Shala: Vaidya Shala to Vaiydya shala-7
57. Ganita Sutras Initiations-1
58. Ganita Sutras Initiations-2
59. Ganita Sutras Initiations-3
60. Ganita Sutras Initiations-4
61. Ganita Sutras Initiations-5
62. Ganita Sutras Initiations-6
63. Ganita Sutras Initiations-7
64. Hyper Cube-5 as Ganita Sutra's Organization Format-1
65. Hyper Cube-5 as Ganita Sutra's Organization Format-2
66. Hyper Cube-5 as Ganita Sutra's Organization Format-3
67. Hyper Cube-5 as Ganita Sutra's Organization Format-4
68. Hyper Cube-5 as Ganita Sutra's Organization Format-5
69. Hyper Cube-5 as Ganita Sutra's Organization Format-6
70. Hyper Cube-5 as Ganita Sutra's Organization Format-7
71. Approach focus: Ganita Sutras Along artifices of Numbers-1
72. Approach focus: Ganita Sutras Along artifices of Numbers-2
73. Approach focus: Ganita Sutras Along artifices of Numbers-3
74. Approach focus: Ganita Sutras Along artifices of Numbers-4
75. Approach focus: Ganita Sutras Along artifices of Numbers-5
76. Approach focus: Ganita Sutras Along artifices of Numbers-6
77. Approach focus: Ganita Sutras Along artifices of Numbers-7
78. Approach focus: Second element/Jal Tatav/Water element-Aspect-1
79. Approach focus: Second element/Jal Tatav/Water element-Aspect-2

80. Approach focus: Second element/Jal Tatav/Water element-Aspect-3
81. Approach focus: Second element/Jal Tatav/Water element-Aspect-4
82. Approach focus: Second element/Jal Tatav/Water element-Aspect-5
83. Approach focus: Second element/Jal Tatav/Water element-Aspect-6
84. Approach focus: Second element/Jal Tatav/Water element-Aspect-7
85. Mathematics of Two-1
86. Mathematics of Two-2
87. Mathematics of Two-3
88. Mathematics of Two-4
89. Mathematics of Two-5
90. Mathematics of Two-6
91. Mathematics of Two-7
92. Mathematics of Two-8
93. Mathematics of Two-9
94. Mathematics of Two-10
95. Mathematics of Two-11
96. Mathematics of Two-12
97. Vedic Mathematics Answer-1
98. Vedic Mathematics Answer-2
99. Vedic Mathematics Answer-3
100. Vedic Mathematics Answer-4
101. Vedic Mathematics Answer-5
102. Vedic Mathematics Answer-6
103. Vedic Mathematics Answer-7
104. Vedic Mathematics Answer-8
105. Vedic Mathematics Answer-9

106. Vedic Mathematics Answer-10
107. Vedic Mathematics Answer-11
108. Vedic Mathematics Answer-12
109. Vedic Mathematics Answer-13
110. Vedic Mathematics Answer-14
111. Vedic Mathematics Answer-15 & 16
112. Vedic Mathematics Answer-16 & 17

*****

Appendix-III

Here below is the list of 84 Lesson of free VM Course No. 3 uploaded on the site (www.learn-and-teach-vedic-mathematics.com)

Chase Steps – Mathematical Chase of Sanskrit

No	Title of the Chase Step
1.	Forty two letters coordinated by 14 Maheshwara Sutras
2.	Fifty two letters Alphabet format
3.	Ativahkas Carry The Transcendental World as Middle Space
4.	Fourth Fold Script Forms on the format of Outer Most Fold of Parnavah
5.	Third Fold Script Forms on The Format of The Fold Prior to The Outer Most Fold of Parnavah
6.	Second Fold Script Forms on The Format of The Second Fold of Parnavah
7.	First Fold Script Forms on The Format of Inner Most Fold Of Parnavah
8.	Vowels: Eighteen folds of first Vowel
9.	Vowels: Parallel Geometric Format for Eighteen Folds of Akara
10.	Vowels: Manifestation Layer (3,4,5,6) and Eighteen Folds of Akara
11.	Vowels: One Hundred and Thirty-two
12.	Vowels: First Maheshwara Sutra
13.	Need of Learning through A Teacher as to How to Pronounce Sounds of Sanskrit Letters
14.	Panini's Shiksha
15.	Sandhi of Vowels-1
16.	Sandhi of Vowels-2
17.	Sandhi of Vowels-3

18. Sandhi of Vowels-4
19. Sandhi of Vowels-5
20. Sandhi of Vowels-6
21. Sandhi of Vowels-7
22. Organization format of Devnagri Alphabet-1
23. Organization format of Devnagri Alphabet-2
24. Organization format of Devnagri Alphabet-3
25. Organization format of Devnagri Alphabet-4
26. Organization format of Devnagri Alphabet-5
27. Organization format of Devnagri Alphabet-6
28. Organization format of Devnagri Alphabet-7
29. Approach focus: Geometric formats of Sanskrit Grammar-1
30. Approach focus: Geometric formats of Sanskrit Grammar-2
31. Approach focus: Geometric formats of Sanskrit Grammar-3
32. Approach focus: Geometric formats of Sanskrit Grammar-4
33. Approach focus: Geometric formats of Sanskrit Grammar-5
34. Approach focus: Geometric formats of Sanskrit Grammar-6
35. Approach focus: Geometric formats of Sanskrit Grammar-7
36. Maheshwara Sutras Approach focus-1
37. Maheshwara Sutras Approach focus-2
38. Maheshwara Sutras Approach focus-3
39. Maheshwara Sutras Approach focus-4
40. Maheshwara Sutras Approach focus-5
41. Maheshwara Sutras Approach focus-6
42. Maheshwara Sutras Approach focus-7
43. Saraswati Mantras Approach focus-1
44. Saraswati Mantras Approach focus-2
45. Saraswati Mantras Approach focus-3
46. Saraswati Mantras Approach focus-4
47. Saraswati Mantras Approach focus-5

48. Saraswati Mantras Approach focus-6
49. Saraswati Mantras Approach focus-7
50. Asthadhyay Text Chapter-1
51. Asthadhyay Text Chapter-2
52. Asthadhyay Text Chapter-3
53. Asthadhyay Text Chapter-4
54. Asthadhyay Text Chapter-5
55. Asthadhyay Text Chapter-6
56. Asthadhyay Text Chapter-7
57. Scripture formats-1
58. Scripture formats-2
59. Scripture formats-3
60. Scripture formats-4
61. Scripture formats-5
62. Scripture formats-6
63. Scripture formats-7
64. Scripture formats-8
65. Scripture formats-9
66. Scripture formats-10
67. Scripture formats-11
68. Scripture formats-12
69. Scripture formats-13
70. Scripture formats-14
71. Scripture formats-15
72. Scripture formats-16
73. Scripture formats-17
74. Scripture formats-18
75. Scripture formats-19
76. Scripture formats-20
77. Scripture formats-21

78. Scripture formats-22
79. Scripture formats-23
80. Scripture formats-24
81. Scripture formats-25 & 26
82. Scripture formats-27 & 28
83. Scripture formats-28, 29 & 30
84. Scripture formats-31, 32 &33

*****

Appendix-IV

Here below is the list of 84 Lesson of free VM Course No. IV uploaded on the site (www.learn-and-teach-vedic-mathematics.com):

Flow Progress Stages – Transcendental Basis of Human Frame

No	Title of the Flow Progress Stage
1.	Sathpatya Measuring Rod
2.	Shad Chakra format
3.	Within Creator's Space
4.	Ten Transcendental Circuits of Human Frame
5.	Fourth and Fifth Vowels
6.	Consciousness State
7.	Unity State of Consciousness
8.	Shaririko Upanishad
9.	Shaririko Upanishad: Aspect-1
10.	Shaririko Upanishad: Aspect-2
11.	Shaririko Upanishad: Aspect-3
12.	Shaririko Upanishad: Aspect-4
13.	Shaririko Upanishad: Aspect-5
14.	Shaririko Upanishad: Aspect-6
15.	Gorakshko Upanishad-1
16.	Gorakshko Upanishad: Updesha-1
17.	Goraksho Upanishad: Updesha-2
18.	Gorakshko Upanishad: Updesha-3
19.	Gorakshko Upanishad: Updesha-4
20.	Gorakshko Upanishad: Updesha-5
21.	Gorakshko Upanishad: Updesha-6
22.	Self-Referral Sustenance-1

23. Self-Referral Sustenance-2
24. Self-Referral Sustenance Prashno Upanishad-2
25. Self-Referral Sustenance Prashno Upanishad-3
26. Self-Referral Sustenance Prashno Upanishad-4
27. Self-Referral Sustenance Prashno Upanishad-5
28. Self-Referral Sustenance Prashno Upanishad-6
29. Approach focus Braham: Sathul sharir & Sathul Sharir-Suksham sharir-1
30. Approach focus Braham: Sathul sharir & Sathul Sharir-Suksham sharir-2
31. Approach focus Braham: Sathul sharir & Sathul Sharir-Suksham sharir-3
32. Approach focus Braham: Sathul sharir & Sathul Sharir-Suksham sharir-4
33. Approach focus Braham: Sathul sharir & Sathul Sharir-Suksham sharir-5
34. Approach focus Braham: Sathul sharir & Sathul Sharir-Suksham sharir-6
35. Approach focus Braham: Sathul sharir & Sathul Sharir-Suksham sharir-7
36. Approach focus: Suksham Sharir-States of consciousness-1
37. Approach focus: Suksham Sharir-States of consciousness-2
38. Approach focus: Suksham Sharir-States of consciousness-3
39. Approach focus: Suksham Sharir-States of consciousness-4
40. Approach focus: Suksham Sharir-States of consciousness-5
41. Approach focus: Suksham Sharir-States of consciousness-6
42. Approach focus: Suksham Sharir-States of consciousness-7
43. Approach focus: Karan Sharir-1
44. Approach focus: Karan Sharir-2
45. Approach focus: Karan Sharir-3
46. Approach focus: Karan Sharir-4
47. Approach focus: Karan Sharir-5

48. Approach focus: Karan Sharir-6
49. Approach focus: Karan Sharir-7
50. Approach focus: Pursha format-1
51. Katha Upanishad
52. Aiteriya Upanishad
53. Taitriya Upanishad
54. Swateshwara Upanishad
55. Patanjali Yogdarshan
56. Atamprabodho Upanishad
57. Vedic Geometry at base of human frame-1
58. Vedic Geometry at base of human frame-2
59. Vedic Geometry at base of human frame-3
60. Vedic Geometry at base of human frame-4
61. Vedic Geometry at base of human frame-5
62. Vedic Geometry at base of human frame-6
63. Vedic Geometry at base of human frame-7
64. Vedic Geometry at base of human frame-8
65. Vedic Geometry at base of human frame-9
66. Vedic Geometry at base of human frame-10
67. Vedic Geometry at base of human frame-11
68. Vedic Geometry at base of human frame-12
69. Vedic Geometry at base of human frame-13
70. Vedic Geometry at base of human frame-14
71. Vedic Geometry at base of human frame-15
72. Vedic Geometry at base of human frame-16
73. Vedic Geometry at base of human frame-17
74. Vedic Geometry at base of human frame-18
75. Vedic Geometry at base of human frame-19
76. Vedic Geometry at base of human frame-20
77. Vedic Geometry at base of human frame-21

*****

Appendix-V

Transcendental Basis of Vedic Mathematics

(Panel of Jury set up by the Bhartiya Vidya Bhavan to select candidates for Shri Guru Gangeshwaranandaji Veda Ratna Puraskar-1997 had selected Dr. S.K. Kapoor for his excellence and service rendered in the field of Vedic Mathematics. The Panel of Jury had considered the evaluation report titled "Transcendental Basis of Vedic Mathematics". As good information about some aspects of Vedic mathematics of the report would be of great interest for the research scholars working in the field and as such same is being included here).

My **studies** culminating into the Doctoral Degree for my thesis titled "Mathematical Basis of Vedic Literature" accepted by Faculty of Mathematics, Kurukshetra University, Kurukshetra, summed up conclusions as under:

Sum up of the results as those flow while working florescence-systems on Om formulation within framed domains are:

1. In a nutshell, the conclusions are that the mathematics precedes the composition of Rigveda Samhita. Vedic Mathematics helped to transform the Universal set of knowledge as a speaking language and in the process itself as well stood transformed as such and assimilated its identity into the Vedas. Within Vedas, all disciplines of knowledge transform their identity and get assimilated into florescence systems admitting 432000 components channelizable on Om formulation. As such, need is to rewrite the history of mathematics, even the intellectual history of man well admitting fundamental unity of human intellect, the message with which the Rigveda ends "Samanam-astu......... Sushasati".
2. One line conclusion, of all the above conclusions, is that the entire Lower Vedic Mathematical domain stands exhaustively covered by single formula:

3. And the above formulation as well is nothing but the extended version of Divya Ganga flow through Om formulation. Divya Ganga flow crystallizes as florescent frames for the Eternal paths beyond and to which the Vedic florescence systems ultimately transformed.

The thesis is divided into 12 chapters with contents as:

1. Vedic mathematics
(Introductory concepts)
– To settle the text

2. Om formulation
(Vedic mathematics applications)
– To settle equivalent formulations:

- Om
- Pranava
- Aum
- Onkar
- Udgitha
- Vashatkara
- Swastik
- Anmoh

3. Vedas
(Mathematics as a speaking language)

- Rigveda
- Yajurveda
- Samaveda
- Atharvveda

4. Adi Valmiki Ramayana
(Vedic mathematical basis)
– Derivation of the frames and systems on
Om formulation in terms of framed domains sequence.

5. Urmahabhartam
(Vedic mathematical basis)
– Derivation of the frames and systems on
Om formulation in terms of framed domains sequence.

6. **Srimad Bhagwad Gita**
   (Vedic mathematical basis)
   – Derivation of the frames and systems on
   Om formulation in terms of framed domains sequence.

7. **Sri Vishnu Shastranam Satotram**
   (Vedic mathematical basis)
   – Derivation of the frames and systems on
   Om formulation in terms of framed domains sequence.

8. **Srimad Durga Saptsati**
   (Vedic mathematical basis)
   – Derivation of the frames and systems on
   Om formulation in terms of framed domains sequence.

9. **Upanishads**
   (Vedic mathematical basis)
   – Derivation of the frames and systems on
   Om formulation in terms of framed domains sequence.
   – Prasno Upanishad; Derivation of the frame and systems.

10. **Shad Darshan**
    (Vedic mathematical basis)
    – Derivation of the frames and systems on
    Om formulation in terms of framed domains sequence of
    - Nyaya Darshan
    - Sankhya
    - Vaisesika
    - Yoga
    - Mimamsa
    - Vedanta
    - Sankarshna

11. **Puranas**
    (Vedic mathematical basis)
    – Derivation of the frames and systems on
    Om formulation in terms of framed domains sequence of Puranas.
    – Srimad Bhagwat Puran; Derivation of frame & system.

12. Devanagri alphabet

(Vedic mathematical basis)

– Derivation of frame and systems of the alphabet, its structure, forms and formulations of the letters within Pranava domain on

Om formulation for working out the sunlight florescence systems within framed domains.

- Maheshwara Sutras and Sanskrit grammar.
- Saraswati Mantras and Vedic grammar.
- Derivation of classical and languages alphabet.
- Mother computer language for all computer systems.

With respect to the value of research work, Professor in Structure Engineering, **Dr. Anand Parkash** of Civil Engineering Department, University of Roorkee, had conveyed in 1986 to Prof. L.R. Varmani, Department of Mathematics in Kurukshetra, Kurukshetra:

"It is my pleasure to convey to you my appreciation of the work undertaken by Mr. S.K. Kapoor in the basis artifice of our alphabet and number system conveying the basis energies channelization into atomic frames and their structures. His study of the sunlight reflecting on the different number frames resulting into all artificial intelligence and value judgment has the potential of a breakthrough in scientific thoughts bearing on the understanding of our universe permeating throughout our vedic literature.

I take this opportunity to congratulate you all the all-valuable inspiration and guidance for the brilliant research output by him.

With regards."

Simultaneously, **Professor Anand Parkash** had enclosed general appreciation of the research result as under:

"It was very satisfying to listen Mr. Sant Kumar Kapoor to decipher the hidden code in the construction of alphabets and numbers channelising the basis energies in the atomic frames constituting their structure. His formulation that there are 108 basic energy sources constituting the fundamental format of the atomic tables is noteworthy.

His attempt to have breakthrough with the insight into sunlight being the basic source of all artificial intelligence and value judgment in this universal, interlinking the modern scientific knowledge with the vedic thoughts, deserve all encouragement and support for advancing the cause of applied research.

I with him all success in his mission."

Professor J. N. Kapur, a well-known mathematician of our country, having hundreds of research articles and dozens of books of mathematics, and experience of more than four decades of actual teaching of mathematics at graduate and postgraduate levels, and of guiding several research scholars, had occasion to consider the geometrical validity of the framed domain sequence being arrived at in the thesis i.e. (a^n : $2na^{n-1}$, n = 1, 2, 3, 4, ...) and his considered opinion conveyed to Professor Krishanaji, read as:

"The sequential interpretation of $a^3/6a^2$, $a^4/8a^3$, $a^5/10a^4$ are valid.

'n' dimensional hypercube has 'n' dimensional content 'a^n'. It has 'n–1' dimensional content '$2na^{n-1}$' and the ratio is '$a^n/2na^{n-1}$'. This has also 'n–2' dimension, 'n–3' dimension, ... content which I have discussed in my article.

This part can be rigorously justified."

Professor Krishnaji in his letter dated March 31, 1988 addressed to Hon'ble Chief Justice of Punjab & Haryana High Court had requested for sparing the services of mine for institute of Vedic Science for two years conveying as:

"We must congratulation you for having in your department such a brilliant and creative mind as Sri S.K. Kapoor, C.J.M., Rohtak, who can authentically expound the foundations of all order in nature, as the basis of the Vedic literature available to us through the oral honoured Vedic tradition. The work by him in his Ph.D. thesis is so basic that the authenticity of the Veda and Vedic literature is established with such mathematical precisions and credibility that through his work, it is possible now to establish the traditionally available Vedic literature as the most orderly and comprehensive literature of a perfect science of life. This has the potential to revolutionise the education in all parts of the world and bring perfection to life on Earth.

This revival of the absolute order with which the unified wholeness of natural law functions and maintains orderliness at every stage of evolution of life, is a blessing for every one in this scientific age, which is rising to be the age of enlightenment through Vedic Science, **Maharishi Mahesh Yogi** is currently formulating the Vedic science and will very much appreciate, if Sri S.K. Kapoor's time and talent are available for structuring Vedic science at a rapid pace, which is such a perfect science of life that with its both approaches – subjective and objective, it far exceeds the benefits of modern science which functions through objective approach only. We will be grateful to you, if you kindly spare his services for our Institute of Vedic Science preferably for two years in order to propagate the results of his research to our universities in Europe and America – Maharishi European Research University, Switzerland, Maharishi International University, Norway, Holland and USA and the forthcoming University of Natural Medicine in Brazil and other institutes of higher learning.

With respectful regards" The report of work done during two year (1988 to 1990) as Visiting Professor in the institute with Professor Krishnaji as Chairman, as submitted by Professor Krishnaji to Hon'ble Chief Justice of Punjab and Haryana High Court, Chandigarh, is as under:

"Shri Sant Kumar Kapoor, Chief Judicial Magistrate, Rohtak, was granted leave by your goodself for two years to carry on research in Vedic Mathematics as a Visiting Professor, in our Institute. He joined us on 25th May 1988.

Shri Sant Kumar Kapoor has done exceptionally meritorious work in Vedic mathematics. He has developed concepts involved in the structure of "Om" and the two Vedic structures "Know Brahman quarter by quarter" and "know fourth quarter Brahman as integrated value of the first three quarters." He has also developed Vedic concepts of multi-dimensional geometrical space and structural frames and systems. He has established that the regular bodies of geometrical domains constitute a framed domains sequence.

$a^n/2na^{n-1}$, where n = 1, 2, 3, 4, ...

and a=dimensional unit.

Using the above concepts, he has analysed several Vedic scriptures. Following are the results of his research activity:—

(i) Mathematical concepts and application to structural frames and systems of Sri Sri Vishnu Sahastranam Stotram.

(ii) Vedic mathematical concepts and applications to structural frame and systems of Sama Veda Samhita.

(iii) Vedic mathematical concepts and application to structural frames and systems of sixteen sutras.

(iv) Vedic mathematical concepts and application to structural frames and systems of Shrimad Bhagwad Geeta.

(v) Vedic mathematical concepts and application to unsolved mathematical problems–Five proofs of Fermat's Last Theorem.

All the above have been published as Institute Reports. The last one "Fermat's Last Theorem" has been published in the form of a book. Three Proofs have been published as a research paper in M.I.U. Journal "Modern Science & Vedic Science" Vol.3, No. 1, pp 75-104, 1989 USA. The two proofs have been submitted as a research paper to "Mathematical Reports of the Academy of Sciences" Canada. This work has acquired great importance because the original theorem was proposed by French Attorney and Mathematician Pierre de Fermat in 1637 and nobody has succeeded in getting the proof inspite of continuous attempts by distinguished mathematicians for more than 350 years. Shri S.K. Kapoor has been awarded the Ph.D. degree by the Maharishi Research University, Seelisberg, Switzerland for his work on Sam Ved Samhtia.

The leave of Shri S. K. Kapoor is due to expire on 24th May, 1990, hence this report is sent for your information and official record."

Simultaneously, Professor Krishnaji had made a request with Hon'ble Chief Justice for further leave for Institute, which was allowed and I worked for another two years (1990-1992) as Visiting Professor and my research results have been summed up in one of the articles by Professor Krishnaji in Mathematics Today and reproduced as Appendix A of my book "Fermat's Last Theorem and Higher Spaces Reality Course".

"Vedic mathematics was lost to the modern world over several thousand years. Fortunately, the renowned scholar **Jagadguru Swami Bharti Krisna Tirthahji Maharaj**, Shankrăcharya of Sharda Peeth and Goverdham Math discovered 16 Vedic Sutras and 13 Upsutras in the Parishistha of Atharvved. After continuous Sadhana and hard work, he was able to decode them and get wide ranging mathematical principles and applications from them.

He found that these Sutras cover all aspects of mathematical science in depth. He wrote sixteen volumes on Vedic mathematics covering all aspects of these sutras in depth. Unfortunately, the manuscript was lost which fact was finally confirmed in 1956.

During his last days when his health was failing and his eyesight was weak, he rewrote from his memory an introductory account of the subject. He attained Mahasmadhi in 1960. Later in 1965 a book entitled "Vedic Mathematics" was published. This book is edited by Dr. V.S. Agarwala with the help of Smt. Manula Trivedi on the basis of manuscript and notes left by Swamijee.

This book is therefore, the first book to give a glimpse of Vedic mathematics and to reestablished the subject. The book attracted the attention of several Indians and foreigners. Dr. Narinder Puri of Roorkee University took up the cause of Vedic Mathematics in right earnest.

His Holiness Maharishi Mahesh Yogi spread the message of Vedic mathematics throughout the world by sending Dr. Puri to several countries and making it possible for many schools all over the world start teaching Vedic mathematics.

A former minister of Human Resource Development, in the Government of India, Sh. P.V.Narasimha Rao made a reference in Parliament about the work on Vedic mathematics, which created a wider interest in the nation. Rashtriya Ved Vidya Pratishthan started looking into the matter and appointed a committee on Vedic Mathematics.

During the same period a controversy arose about the source of these 16 Sutras and the level of mathematics contained in them. One view was that these Sutras do not have a Vedic source and that they deal only with simple arithmetical operations.

In the meantime, Dr. Sant Kumar Kapoor, inspired by integrated impact of the Ganita Sutras, started working on basic aspects of the mathematical structure of Vedic literature. He got several clues from Ganita Sutras, Maheshwara Sutra and many other Vedic Mantras.

He was inspired by the divine blessing from Yograj Sri Sripad Babaji of Vrindavan and His Holiness Maharishi Mahesh Yogiji at Maharishi Nagar. He has been awarded Ph.D. degree by Kurukshetra University, in December 1990 on his thesis titled "Mathematical Basis of Vedic literature".

His work shows that the potential of Ganita Sutras is much higher than what appears in the book "Vedic Mathematics" and the mathematical structure of Ganita Sutras is similar to that of Samved indicating that Ganita Stutras belong to Vedic family.

His work shows that the Devnagri alphabet and the Vedic scriptures are mathematically organized. Perhaps a more important aspect of his work in the existence of Real Geometric Spaces of higher dimensions in the form of a continuum.

A comprehensive description of his research is given in the next section. Dr. Sant Kumar Kapoor is presently working as a Visiting Professor in the Indian Institute of Maharishi Vedic Science & Technology, which is a part of Maharishi Ved Vigyan Vishwa Vidyapeeth, Maharishi Nagar. Three research students are working in our institution on the structure and properties of fourth, fifth and sixth dimensional spaces.

Vedic Mathematical Research done by Dr. S. K. Kapoor

Studies reveal that knowledge and organisation of knowledge are two distinct disciplines. The thesis of Dr. S.K.Kapoor is substantively covering the organisational aspect of Vedic knowledge. The main results of the thesis and his subsequent research can be summed as under:

1. *Real 4 and higher dimensional spaces do exist.*

Real 4 and higher dimensional spaces were not only known to the Vedic seers but those were precisely availed by them for organisation of Vedic knowledge.

Vedic alphabet format is a precise setup of 4-space. The alphabet letters of Devnagri script, their forms formulations, combinations and rules etc. are precisely mathematical.

Sakala Rigvedic Samhita is the oldest scripture of mankind. The organisational format of this Samhita is precisely that of real 6-space.

Vedic knowledge accepts a measuring rod constituted by regular bodies of the first six real dimensional spaces which admit formulations as:

Interval	Square	Cube	Hypercube-4	Hypercube-5	Hypercube-6
1-space body	2-space body	3-space body	4-space body	5-space body	6-space body

The organisational format of Srimad Bhagwad Gita emerges to be of prime importance as it is parallel to the organising power of the knowledge content of the scripture. This format tallies with the format of human body on the one hand and the sun on the other hand as of real 6-space.

The Trinity of Gods namely Brahma, Mahesh, Vishnu are the overlords of real 4, 5 and 6 spaces respectively.

The geometrical continuum expressed as manifestation layers of 4 folds of consecutive dimensional spaces contents. These folds of the nth manifestation layer can be represented as under:

First Fold	**Second Fold**	**Third Fold**	**Fourth Fold**
Dimension	Frame	Domain	Origin
(n-2) space content	(n-1) space content	n-space content	(n+1) space content

The transcendence from one manifestation layer to another manifestation layer giving rise to the following (five steps) chain reaction or five steps, which are possible within the setup of Panch Mahabhut.

Manifestation Layer	**Dimension**	**Frame**	**Domain**	**Origin**
nth	Space fold (n-2)	Space fold (n-1)	Space fold n	Space fold (n+1)
(n+1) nth	Space fold (n-1)	Space fold n	Space fold (n+1)	Space fold (n+2)

Transition from one space to another space is to be had in terms of unlocking of the seals of the origin points of all the four folds of the manifestation. The modern mathematical models of transition from straight line to plane deserve serious reexamination. In particular the axioms of space filling curves and the axioms of 'one' without a predecessor deserve close scrutiny as their rationale emerges to be without basis.

The role of real numbers additive group (R,+) and real numbers field (R,+,×) with reference to straight line deserve to be differentiated.

The plane deserves to be studied as four geometrically distinct quarters. One faced plane and two-faced plane are two distinct geometrical setups and they deserve to be taken up as such.

The concepts of origin and dimension are two concepts with respect to which the modern geometrical models are not up to date. These two concepts deserves to be studied in detail as transcendence to the higher dimensional spaces is possible only in terms of their understanding.

Human body is a compactified phenomenon of multi-layer physiological existence. The start with state of existence is that of waking state, which is parallel to the expression of 1-space as dimension into 3-space domain. Sequentially, the existence phenomenon unfolds until seventh state of consciousness, which would be corresponding to the 7-space as dimension into 9-space domain.

The origin point of the 6-space, being the 7-space setup, the human body, geometrically, turns out to be hypercube-6 and this would explain how the primordial sound, the planetary effects, the Yajna oblations etc. operate and precisely influence the individual existence patterns.

Srimad Bhagwad Gita is one such scripture whose organisational format precisely workout for us the structural set up and frames of the 6-space. The study zone of Srimad Bhagwad Gita can be worked as under:

Srimad Bhagwad Gita
Study Zone

$a/2$	$a^2/4a$	$a^3/6a^2$	$a^4/8a^3$	$a^5/10a^4$ $a^6/12a^5$ $a^5/10a^4$	$a^4/8a^3$	$a^3/6a^2$	$a^2/4a$	$a/2$
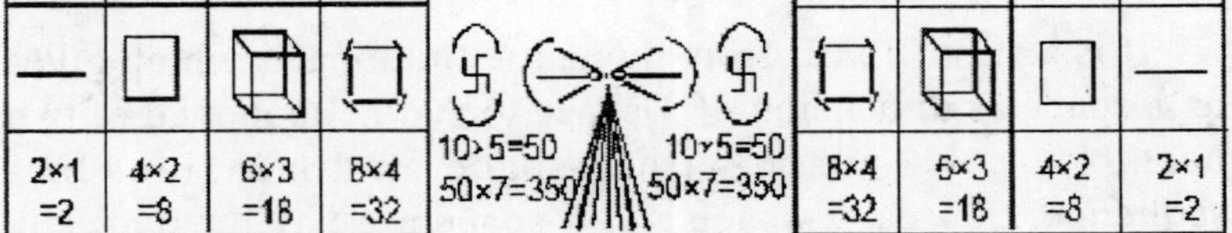								
2×1 =2	4×2 =8	6×3 =18	8×4 =32	10×5=50 50×7=350 10×5=50 50×7=350	8×4 =32	6×3 =18	4×2 =8	2×1 =2
Orbitals				350+350=700				
2	6	10	14	18=5+6+7				

The organisational setups of Ganita Sutras, Maheshwara Sutras, Saraswati Mantras, Gyatri Mantra and Om formulation deserve interdisciplinary explorations.

Sankhay Nistha and Yoga Nistha are complementary and supplementary of each other and as such their complementary nature and supplementary nature deserve to be distinguished well. Non-differentiation of the same is bound to deprive us of most of the results in specific forms.

Now let me take up how the subject contents of my different books are indexed in their prefaces. First starting with "Vedic Mathematical Study of Structural Frames and Systems of Sri Sri Vishnu Sahastranam Stotram", published by Maharishi Ved Vigyan Vishwa Vidyapeetham, Maharishi Nagar–201304, (INDIA) (1991):

"Preface

The present study aims at approaching the organisation format of the scripture; Sri Sri Vishnu Sahastranam Stotram. This is a scripture of 107 verses and it composes 1000 names of Lord Vishnu, who is overlord of real 6-space.

Though the modern mathematics too talks of dimensional spaces higher than three but those are mere mental-constructs. The modern mathematics has striped the concepts of dimension from geometrical reality and as such the suffix 'Real' before 6-space is of real significance here in this study.

If it is only while faced with problems of the classifications of the geometries of dimensional spaces and the sporadic finite groups, we are really face to face with the intensity and richness of each space beyond 3 space. Sylvestor's classification theory of n+1 geometries of n-space is just a property of real 4-space.

It is a simple shift from monad without parts which gives us 1-space as dimension of 3-space to synthetic monad of two parts of real 4 space with 2-space as dimension. In a sequence, the real 5-space and 6-space are having synthetic monads of 3 parts and 4 parts respectively and accordingly their dimensions are 3-space and 4-space.

Vedic literature avails real 4, 5 and 6 spaces for organisation of knowledge. These spaces respectively contribute 24, 25 and 26 Tatavs (elements/characteristics properties).

The step beyond Sylvester's classification or n+1 geometries of n-space would be to enter real 5-space where we shall be getting 2n+1 geometries of signatures –n to +n of n-space.

The step beyond the 25th sporautic finite group would be a step into the real 6-space and we straight way get the number of elements of 26th sporadic finite group as 10×12^{64}. And with it, the classification assignment over which practically we have consumed complete nine decades or this century and still 26th group is eluding, gets answer from ancient wisdom of real spaces.

This would help us appreciate the ancient wisdom to accept the measuring rod of real bodies of first six real spaces admitting ratios of their domain parts as $a^n : 2na^{n-1}$, n = 1, 2, 3, 4, 5, 6.

This all is being added here just to impress upon the rationale for the choice of Sri Sri Vishnu Sahastranam Stotram for organisational study.

However, I would like to add at the outset that by all means it is just a preliminary study into the structural frames and systems of this scripture and the real studies would follow from the pens of really the real privileged souls who, I pray, should spare moments from their blissful samadhi to give the desired direction to the flow of intelligence of the present generation.

Sri Sri Vishnu Sahastram Stotram is the holiest of the Holy Scriptures whose enlightenment was the last discourse to the Pandavas given by great Bishma from his deathbed of arrows in the battlefield of Kurukshetra in the presence of Lord Krishna, incarnation of Lord Vishnu.This, that way, is the scripture of perfection and hence the proper scripture of the Sankhya nistha as well as the Yoga nishta.

His Holiness Sri Sripad Babaji initiated me into Sankhya nistha as the science of organisation of the Universe on the artifice of natural numbers. His Holiness Maharishi Mahesh Yogi has initiated me further in the subject that the Sankhya nistha is the science of self-organising power of the pure knowledge.

The specific processing lines, which I describe as Sripad processing line is to know Brahman quarter by quarter and the Maharishi processing line is to know fourth quarter of Brahman as integrated quarter of first three quarters.

The present study in a way is nothing but the extent to which I could comprehend only that much and over and above that are the limitations on language and what not and so the humble request that they may skip over such stages by their own contemplation which is the only guide in such like adventures or mortals like me regarding the vedic scripture of perfection.

× × ×

With all humbleness this study is submitted with the hope that the advanced students of Vedic science and technology will continue the processing further into this and other scriptures. Here is a caution for them that the first correction which the modern mathematics needs in the discipline of geometry is regarding its hypothesis of space filling curves.

This hypothesis deserves to be rejected. The second thing which I would like to suggest is that the continuum be approached the Vedic way as manifested layers of four-folds of four consecutive space-contents and not 'individual dimensional space wise' as is being attempted at present by the modern mathematics.

Before submitting the present study for consideration of the scholars I feel it may duty to share the first principles of decoding the knowledge content from the organisational format of the Vedic scriptures. The keyword of this principle is Shardha and culmination is the word nistha.

The principle operates as: Samsayatma Vinasyanti (Gita-4.40). Practical working rule comes to be that the doubt destroys everything, the scriptural commands deserves to be approached faithfully, the words of seers deserve to be taken up on their face value.

To have an idea that how difficult it becomes to gain anything from the Vedic systems the moment one approaches them with an initial doubts, we can see what mess stands created by the individuals about the powerful systems of "Ganita Sutras."

Professor Krishnaji has expressed by way of Forword of this book as:

"I am happy that, I am able to seek the blessings of H.H Maharishi Mahesh Yogi on the first publication of the Indian Institute of Maharishi Vedic Science & Technology in a record time. Shri S.K. Kapoor, who is Visiting Professor in the Institute has produced, wonderful work, entitled "Vedic mathematical concepts and its application to structures and systems of Sri Sri Vishnu Sahastranam Stotram".

This publication is the first publication or its kind in which Sh. S.K.Kapoor has introduced the Vedic geometric formula called "Gurudev formula" and has introduced the concepts of "Divya Ganga flow" and "Om formulation".

He has explained the basic difference between conventional geometry and Vedic geometry e.g., the point in

Vedic geometry has a structure whereas the point in conventional geometry is dimensionless.

He has gone into sufficient yet restrained details of the new concepts and the rules before applying them to the structural systems or the scripture Sri Sri Vishnu Sahastranam Stotram.

He has shown without any strain and without any doubt that Vedic mathematical concepts are able to explain unambiguously, the number of names of Lord Vishnu included in every individual Shaloka of the scripture.

His writing comes directly from his thinking. His gift of thinking is original but the inspiration given by H. H. Shri Pad Babaji of Vrindavan and H. H. Maharishi Mahesh Yogi is clearly visible. The knowledge is drawn from the vast treasure of Maharishi Vedic science and Technology but the representation is original.

In recent year, there has been a revival of the ancient Vedic knowledge because a very large section of people have started realising that Vedas are the infinite source of complete knowledge and have existed from times immemorial.

However, there has been an unfortunate tendency, primarily due to ignorance, of comparing and trying to understand Vedic science in terms of the modern science. Very few people realise that the modern science utilises only the objective approach whereas Maharishi Vedic Science utilises both objective and subjective approaches to gaining knowledge simultaneously.

The epistemological nature of the ancient Indian knowledge is based on the understanding that the process of intuitive revelation of true knowledge in the individual mind can be controlled by eliminating the difference in the operation of emotions and intellect through their unification by the techniques of Yoga Darshan.

It will be of interest to the student of Vedic science, if they carefully understand the difference between modern science and Maharishi Vedic science.

Vedic mathematics is one particular component of Vedic science. Mathematics is called the queen of sciences and is the fundamental base on which all planning is done and all structures are raised.

Shri S.K. Kapoor is using this as an axiom and has applied the Vedic mathematics concepts as an integral part of Maharishi Vedic Science and Technology to investigate in the volume a particular case of structures and systems of Sri Sri Vishnu Sahastranam Stotram.

The readers may find it difficult to comprehend the subject during the first reading because they are used conditioned to the usual objective approach of conventional mathematics.

The second and subsequent readings, however, will reveal to them the real meaning of concepts and therefore, the immense beauty and power of this knowledge.

The Indian Institute of Maharishi Vedic Science and Technology is a research institute covering all aspects of Vedic Science & Technology including Vedic philosophy, structure of Vedic knowledge, Vedic mathematics, Ayurveda, Jyotish, Ganderva Veda, Dhanurveda, Sathapatyaveda etc.

The Vedic science covers the widest possible spectrum from point to infinity and back to the point and the infinitely expanding universe. I hope that the present publication will re-open the infinite sources to pure knowledge, which has always existed in the supreme consciousness and has been frozen in the self-referral individual consciousness also from times immemorial.

One will find that before the main text there is an abstract table giving geometrical coordination of 1000 names of Lord Vishnu. To begin with, it may be incomprehensible to the understand reader, but it will certainly show that the Vedic geometry deals with the dimensions higher than 3, that is 4th, 5th, 6th dimensional frames whereas conventional geometry does not go beyond third dimensional frame.

The implications of this work are multidimensional and far-reaching. One aspect is the Vedic claim that faithful recitation of Sri Sri Vishnu Sahastranam Stotram can cure all kinds of fevers. This can easily be understood because the studies reveal:

"The Vedic sounds are multi-dimensional domains' frequencies from within a particular dimensional domain as the structure of that domain. When the sounds are pronounced, the frozen frequencies get initiated and the self-organising power of the Vedic sounds sets the frequencies' potentialisation process into action. It is this process whose utilisation is the aim of different Vedic scriptures."

This work is being submitted at the feet of H.H. Maharishi Mahesh Yogi for his blessing.

Jai Guru Dev!"

The book "Vedic Geometry" of mine published by M/s. Arya Book Depot, Karol Bagh, New Delhi (in 1994) indexes it preface as:

"The present introductory volume aims to introduce the ancient discipline of Vedic geometry. We may formally define Vedic geometry as a discipline of Geometry based on Vedic concepts.

Studies of mathematical basis of Vedic literature reveals that 4 and higher dimensional reality was not only known to Vedic Rishis but also put to practical use by them for organisation of pure knowledge on geometric formats, particularly, real 4, 5 and 6 space formats.

Vedic comprehension of geometric reality is characteristically different from our present-day conclusion. Basic difference in the two approaches viz. modern approach vis-a-vis Vedic approach can be appreciated in terms of the concept of dimension. Modern Cartesian dimensional approach makes all dimensional spaces the linear dimensional spaces as much as that 1-space is to play the role of dimension for all dimensional spaces while, on the contrary,

Vedic comprehension comes to be that n-2 space plays the role of dimension of n-space.

This glaring difference may well be focused algebraically as that as per modern model we may express 4, 5 and 6 spaces as A^4, A^5 and A^6 while as per Vedic model their expressions comes to be $(A^2)^4$, $(A^3)^5$ and $((A^2)^4)^6$.

The above difference and impact of n–2 space playing the role of dimension of n-space comes to be that it requires 4 consecutive dimensional spaces to manifest a dimensional body as a 4 fold body having distinct dimension fold, boundary fold, domain fold and origin fold.

Our well-known solid bodies having cubes as their representative regular bodies avail 1-space as dimension fold, 2-space as boundary fold, 3-space as dimension fold. and 4-space as origin fold. In general n–2, n–1, n and n+1 spaces together synthesis and manifest 'n' dimensional body which may be designated as n–2 manifestation layer.

Also it may be designated as hypercube-n. This makes our modern set theory or binary compositions a study of 0 manifestation layer and Euclidean geometry a study of 1-manifestation layer while Vedic geometry takes up all manifestation layers within its domain.

The beauty of this discipline lies in its capacity to unify all discipline of knowledge as a single discipline. Illustratively, human body, sun, nucleus or nucleus of atom, 6-space, organisation of knowledge of Srimad Bhagwad Gita etc. etc. are all availing the format of hypercube 6 i.e. 4 manifestation layer.

This would help us appreciate the uniqueness of approach and richness of content of this discipline. But this is just a first glimpse. Real bliss lies ahead with unification of entire existence phenomenon as impulses of consciousness."

This "Vedic Geometry" book has 10 chapters with split-up of contents as under:

Chapter-1

INTRODUCTION OF THE SUBJECT

1.1 Vedic geometry

1.2 Vedic knowledge

1.3 Vedic literature

1.4 Organisation of knowledge

1.5 Geometric formats

1.6 Organisation parallel to knowledge content

1.7 Four space

1.8 Five space

1.9 Six space

1.10 Trinity of Gods.

Chapter-2

BASIC CONCEPTS AND FORMULATIONS

2.1 Synthetic monads

2.2 Manifestation layers

2.3 Transcendence at the origin

2.4 Regulation acceleration of transcendence

2.5 Measuring rod

2.6 Generation status of elements

2.7 Place value system

2.8 Power sets

2.9 Reflection operation

2.10 Structural constants

Chapter-3

TRINITY OF GODS

Chapter-4

SHADCHAKRAS

Chapter-5

VEDIC GEOMETRY

Chapter-6

VEDIC NUMBER THEORY

9.3 Organisation of the text of Srimad Bhagwad Gita

9.4 Organisation of the knowledge of 18 chapters.

Chapter-10

CONCLUSION AND THEIR APPLICATIONS TO THE SOLUTION OF FERMAT'S LAST THEOREM

10.1 Fermat's Last Theorem

10.2 Four folds approach

10.3 First proof

10.4 Proof by direct comparison

10.5 Transcendence approach

10.6 Power expression

10.7 Power sets

10.8 Aspects 1 to 12.

International Sanskrit Research Academy (Regd. Trust), Bangalore in its letter dated June 22, 1996, conveyed its editorial team's view regarding Chapter-9 titled "Structural frames and systems of Srimad Bhagwad Gita" of my book "Vedic Geometry" (pages 433 to 458) as under:

"This is an inter disciplinary study bringing out the link between the Vedic geometry and the classical text of philosophy namely Bhagavadgita in the Chapter-9. This study can further lead us to understand the relation of Gita with other important scientific disciplines in the Indian tradition about which the present knowledge is very meager. This article is a stimulator for further studies."

In the Preface to my book "Fermat's Last Theorem and Higher Spaces Reality Course" published by M/s. Arya Book Depot, the subject content is indexed and focused as under:

"Preface

Present phase of intellectual history of man begins with seventeenth century. Over all these years, one statement, which remained source of intellectual challenge, is the statement of Fermat's Last Theorem. It is rightly stated "those non-mathematicians who really wants to understand the essential nature of mathematics and how it is created ought to begin with this discussion" (of our recorded history how bravely we fought to conquer the fort of this statement).

The French Attorney and mathematician Pierre de Fermat (1601–65) made his assertion in 1637 as the following note (original in Latin admitting English rendering translation as given below) in the margin of mathematics book (Bachet's translation of Diophantus' Arithmetica, great classics of ancient Greek mathematics) he was reading at that time:

It is impossible to separate a cube into two cubes, a biquadrate into two biquadrate, or in general any power higher than second into two powers of like degree. **I have discovered a truly remarkable proof which this margin is too small to contain**'.

Prof. Paulo Ribenboim in his essay titled "Early History of Fermat's Last Theorem" highlights amongst other features, the fact that despite efforts no proof could be traced from the paper left behind by Fermat and that one school of thought is gaining grounds when he adds: 'It is very difficult to understand today how the most distinguished mathematicians could have failed to rediscover a proof if one had existed.'

This article despite above reservations earned a hopeful editorial comment: 'It is amazingly simple question but whose solution eluded the world, the world of mathematics to this very hour.'

The present work approaches this theorem from many angles and put forwards several alternative proofs but the crowning jewel is the equally simple answer for an amazingly simple question. The answer runs as that on format of volume of cube;

$z^3 = x^3 + y^3$ implies

$(z-2)^3 = (x-2)^3 + (y-2)^3$ as

$a^3 = (a-2)^3 + 6(a-2)^2 + 12(a-1)^1 + 12(a-1)^0$

Number value cube = volume + surface area + length of edges + value of corner of geometric cubes.

The law of impossibility of infinite descent proves the case n=3.

As such the dimensional order of odd hypercube-n such that n=2m+1 comes to be 1 × 3 × 5 × 7 ×......... 2m + 1 and for even hypercubes-n such that n=2m comes to be 2 × 4 × 6 × 8 × .... 2m, therefore, for all hypercubes-n such that n=4 implies that dimensional order of hypercube n is divisible by 3. Therefore, n is divisible by 3.

Therefore, the general case for n=4 immediately follows from case n=3 as we can rewrite $z^n = x^n + y^n$ as $(z^{n/3})^3 = (x^{n/3})^3 + (y^{n/3})^3$. The case n=4 was proved by Fermat himself. The initial choice or format of volume of cube restricts n to be greater than 2. Hence, the general proof.

The source concepts at the base of above proof is: n–2 space plays the role of dimension of n-space. It is well preserved in our Vedic literature. Two such references are Uttarkand 79(ka) Ramcharitmanas and Bhagwad Mahapuran (10.9.15-16).

General Proof (at pages 104-110 of this book) was sent by the author to Prof. Paulo Ribenboim who referred it to referee but unfortunately lost it.

To be at most natural answer is an intellectual satisfaction but his is the beginning of the new wonderful higher dimensional reality displayed beyond this point.

The mathematical truth of this theorem is the basic foundation stone for jumping from three space reality to four and higher space reality. Aim of this work is to attempt a school courses for higher dimensional mathematics to mature the young brains by exposing them to the geometric formats of mathematics, science & technology of real three, five and six spaces over a duration of post-middle four years duration of school education so that present generation may have taste of the bliss of Vedic wisdom.

The efforts of the author would get amply rewarded if this work helps the students and teachers of mathematics to think about the ways and means of coming out of a single space (three space) model of mathematics, science and technology.

Eternity of Vedic knowledge is there because of eternal organisation formats of Vedic knowledge. Idols of Trinity of Gods manifest on formats of hypercubes 4, 5 and 6. These together as three folds of four manifestation layer with 4,5,6 and 7 space as dimension, boundary, domain and origin fold constitute a really wonderful format being availed as organisation format for Vedic wisdom.

Because of it, seven-place value system has a unique role to play. As is evident from the table below, all the six numbers 1 to 6 of seven-place value system acquire same value i.e 1 for power 6.

		Digits					
		1	2	3	4	5	6
	1	1	2	3	4	5	6
	2	1	4	2	2	4	1
	3	1	1	6	1	6	6
Powers	4	1	2	4	4	2	1
	5	1	4	5	2	3	6
	6	1	1	1	1	1	1

As such, it may be very easy to physically test even case i.e. n–2m cases for n>4 on seven place value system."

The book "Foundations of Higher Vedic Mathematics" published by M/s. Arya Book Depot, has indexed in its Preface as:

"Ved means 'knowledge'. Studies of available Vedic literature reveal that Vedic seers had successfully organised whole range of knowledge as a single discipline as a speaking language. Further, studies reveal that our ancestors could achieve success par excellence by first comprehending

Reality which ends up as Triloki (three space) as higher dimensional Reality and secondly, by making use of this comprehension of higher dimensional Reality for the required geometric formats for organisation of Vedic knowledge.

Upanishads are the end-product of Vedic literature. The enlightenment of Upanishads takes us to the fundamental unity of Reality as much as that Purusha emerges to be the first manifested incarnation of Brahm (Sri Sri Vishnu Puran). The specific enlightenment of Upanishads is that Lord Vishnu; Sun, Purusha, Atman and real six space are organisationally in unison with each other. It is this enlightened comprehension, which may be taken as the source for unification of knowledge on six-space format.

Further, studies of mathematical basis of literature leads to the conclusion that the organisation format of hypercube-6 (being the representative regular body of six space) is structurally very rich and is capable of organizing the whole range of pure knowledge as a single discipline as a speaking language as real six space, if viewed linearly, is a space of 2×4×6=48 linear dimensional order.

Upanishad further enlighten us that Lord Vishnu accepts hypecube-6 as format for His Idol, Shad-Chakra format of human body admits coordination through Sushmana Nari and takes us to the core of sun parallel to the measuring rod of first six hypercubes accepted by Sthapatya Ved. In fact, every Vedic literature makes use of one or other aspect of hypercube-6 for organisation of knowledge.

In the opening paragraph of chapter-1 of present studies it has been made specific that: Vedic mathematics, as it is, yet to be explored. Though whole domain and the scope of Vedic mathematics, as it is, yet to be explored but three distinct approaches, to which we may have access at present, are, firstly, 'on the artifices of whole numbers' of Ganita Sutras, as interpreted by H.H. Swami Bharti Krishna Tirthaji Maharaj, secondly 'on the geometric formats' of Maheshwara Sutras, as is expounded by H. H. Sri Sripad Babaji Maharaj, and thirdly, 'on consciousness state of human physiology' as unfolded by H.H.Maharishi Mahesh Yogiji Maharaj, known as 'Maharishi's Vedic Mathematics'.

Present studies are titled 'Foundations of higher Vedic mathematics' as here attempt is being made to reach at the foundations of higher Vedic mathematics on which the Vedic knowledge has been organised as a single discipline as a speaking language. Mahalakshmi Astham Stotram has been taken as the reference scripture for measure the success of studies.

Present study for foundations of higher Vedic mathematics is being divided into four chapters as four steps. The first step is to introduce the basic concepts and formats required for reaching at the organisation formats of different folds of Vedic mathematics text. This as such is the topic of study of the first chapter.

The inner evidence of available Vedic literature makes out that Om formulation is the basic formulation without whose proper understanding it may not be possible to have desired insight into the working rules of Vedic mathematics at the base of the organisation of Vedic knowledge and as such the second topic of study of chapter-2 is 'Om Formulation'.

The organisation of knowledge being for working out the systems which end up as Triloki and as such the third topic for detailed study which need be taken and in fact is being taken up in the present study as chapter-3 is 'Structural richness of a cube'.

With Mahalakshmi-Ashtakam Stotram at the centre and transcendental meditation as the basic technology, attempt has been made to ascertain the specific properties of a geometric space as the Santana (eternal format/real 5-space) accepted for the organisation of Vedic knowledge.

Mahalakshmi Ashtakam Stotram is a scripture of Stuti Shalokas devoted to Goddess Mahalakshmi. Devotees by faithfully reciting these Shaloks get blessed with Vibhuti and Shakti of Vishnu-lok. This is the fruit of pure knowledge, the Vedas.

Devotees just recite the Stuti shaloks with full faith and get blessed with the godly grace. For all those who wish to be blessed with such a grace, for them, there is no choice, but to recite these Shalokas with faith. Such readers may straight a way go to the text of the Stotram (chapter-4) and faithfully recite the same be blessed with godly grace of Goddess Mahalakshmi.

The conclusion of the studies comes to be that: Vibhuti and Shakti of the Vishnu-lok manifests at boundary or Vishnu-lok (real 5-space in the role of boundary) as well as Saptrishi-lok as stands released with unlocking of the seal of the origin of Vishnu-lok. This is there because of the Goddess Mahalakshmi, the consort of Lord Vishnu, manifesting as real 5-space playing the role of boundary of 6-space and dimensional order of 7-space.

This privileged state of Vibhuti and Shakti of Vishnu-lok is there because of the Goddess Mahalakshmi. This is the state whose comprehension and understanding has much education for us who are interested in sciences, mathematics and technology of higher real spaces, particularly as to how the continuity of manifestations is being maintained by the dimensional contents which otherwise remain sealed at the origins of the dimensional spaces.

The concept of release of higher dimensional content with unlocking of the seal of the origin and transcendence setting into action is the most fundamental concept, which is not known to the present day sciences and technologies. Four consecutive dimensional spaces manifest simultaneously as four-fold manifestation sequentially playing the roles of dimension, boundary, domain and origin with origin normally remaining in dormant sate but otherwise being capable of being initiative.

This in Vedic technological language would mean chiseling of an eye. In Yogic language it would mean attaining transcendence state for higher state of consciousness. This precisely in geometric language would mean the release of next higher dimensional space.

Though in the present study which aims to outline some of the steps for foundations of higher Vedic mathematics, an attempt is being made to make this study a self-contained one but despite every effort it has not become possible to go for defining each and every technical term used here and as such I feel that one reading of my book on "Vedic Geometry" may be of great help.

Before presenting the studies to the readers, I feel its my duty to express that this study simply would have been without shape, but for the enlightenment of Maharishi's Vedic mathematics' statement sharpened and focused by H.H.Maharishi Mahesh Yogiji Maharaj (as part of Maharishi's Absolute Theory of Defence). This has been annexed with the present studies to give the readers an idea as to what is going to be the final shape of things in the area of Vedic mathematics, and now wonderful worlds of higher dimensional Reality are awaiting for us.

Scholars may straight-a-way go to the Maharishi's Vedic Mathematics statement. The readers, who are not yet exposed to Vedic geometry and Vedic mathematics, may first go through chapter-3 of the present studies regarding "Structural Richness of a Cube". However, the readers who are already exposed to Vedic geometry and Vedic mathematics, they may start with chapter-1 titled "Steps for foundations of higher Vedic mathematics".

The reader who are interested to first to go through Om formulation, they may start with chapter-2 and the readers who are interested only in godly grace they may straight-a-way go to chapter-4 titled "Mahalakshmi-Ashtakam Stotram": Mahalakshmi Ashtakam Stotram blesses devotees reciting the scripture with faith.

With all humbleness, I share with the readers that my family has full faith in this scripture and everyone can be blessed with godly grace just by faithful recitation (three times a day) or Stuti Shaloks of Mahalakshmi Ashtakam Stotram.

Present study, in fact is a joint exercise of all the members of my family as every member has contributed for the completion of the work in his/her own way. Further I would like to share with the readers that the present studies have become possible only because of the privilege to be at the feet to H.H. Sri Sripad Babaji Maharaj and had initiations into the organisation of Vedic knowledge and further being at the feet of H.H. Maharishi Mahesh Yogiji Maharaj and learnt Transcendental Meditation for transcendence values needed for reciting at the inner folds of organisation formats of Vedic knowledge.

Mahalakshmi-Ashtakam Stotram text is a composition of just 818 letter organised as just 11 Shaloks and its recitation even by reading the text is an exercise of less than five minutes. By way of study of chapter-4: "Mahalakshmi-Ashtakam Stotram", it has been expressed about the organisation and composition of this scripture as that: The composition of this scripture as that: The composition range of 818 units is firstly organised as 352 syllables. These 352 syllables are further organised as 85 Padas (words). The Padas range of 85 units is organised as 22 composition ranges as half is organised as 22-composition range as half Shaloks of 16 syllables length. Ultimately these 22 composition ranges get organised as 11 Shaloks of 32 syllables each.

The organisation of individual Shaloks as 32 syllables of equal parts of 16 syllables is parallel to the organisation of 32 teeth in the mouth grouped as two jaws of 16 teeth each.

The human Nad had ultimately gets regulated through these 32 teeth arrangement as Shabad (Mantras). The organisation of Nad of 818 units of Mahalakshmi-Ashtakam Stotram is parallel to the organisation of the affine space of a cube of 818 units. This organisation is further parallel to the organisation range of Vedic mathematics text of precisely 818 units.

Therefore, the organisation of 818 units composition of Mahalakshmi Stotram satisfy the first test of being of the potentialities of the nature which ends up as our Triloki. As such, everything what may be attained within human frame is ensured with faithful recitation of Mahalakshmi-Ashtakam Stotram.

The physical test of the organisation possible on the affine space within a cube and the intellectual test of the organisation of the Vedic mathematics text having been done, the interested readers may further have the satisfaction of the subjective test by themselves experiencing the taste of the Stotram.

For such interested readers, it is added that they should have faithful recitation of the scripture three times a day, initially with the help of written text till the text become fully enlivened in their consciousness, it would become possible for them to dispense with the written text and they may enjoy the recitation of the scripture while sitting in Pada-Asan. Though the recitation as such may turn out to be just of one-minute duration and even stage would reach where just attention would enliven the whole range of the text.

The blessed state of Mahalakshmi-Ashtakam Stotram getting enlivened just the moment there is attention to it becomes a permanent state provided one does not deviate from the laws of pure consciousness state which admits truth as its base/format for enlivened existence.

To achieve the enlivened state of existence may not be difficult but to make it a permanent state, one is to be ever truthful to oneself so that purity of consciousness is not diluted. If at any stage one feels that some circumstances of consciousness are having tendency to dilute the purity of consciousness, he should increase the number of recitations of this scripture to meet those uncalled for tendencies of those circumstances. Faithful recitation has the potentialities to overcome all such things."

This book "Foundations of Higher Vedic Mathematics" published by M/s. Arya Book Depot, has four chapters with split-up contents as:

1. Steps for foundations of Higher Vedic Mathematics

1.1 Introduction

1.2 Main results

1.3 Basic geometric formats and concepts

1.4 Vedic mathematics text

1.5 Illustrative case: Fermat's Last Theorem

1.6 Organisation of Rigved Samhita

1.7 Jyoti Manifestation format

1.8 Triloki & trinity of gods

1.9 Conclusion

2. Om formulation

2.1 Introduction

2.2 First set of rules of Om formulation

2.3 Second set of rules of Om formulation

2.4 Third set of rules of Om formulation

2.5 Fourth set of rules for Om formulation

2.6 Studies: 4 components of Om formulation as first four-dimensional domains

2.7 Illustration I: Structure of an atom

2.8 Illustration II: 120 years cycle of human destiny

2.9 Illustration III: Human physiology: Shadchakra

2.10 Illustration IV: Lagrangian n=8.

2.11 Appendix: Occurrence of 'Om' in the ancient literature and some questions regarding multi-dimensional spaces.

3. Structural richness of cube

3.1 Cube

3.2 Three space

My modest work "Patanjali Unison Discipline" is my first attempt as more in the spirits of a Sadhaka then in that of scholar, was blessed by my Guru His Holiness **Sri Sripad Babaji Maharaj**, with his forrword:

"This Indian spiritual wisdom revealed first in and through the Vedas was well systematised and reconstructed in the six orthodox systems of Indian philosophy, namely

Nayaha, Vaisesika Sankhya, Yoga, Mimasa and Vedanta. Based on scriptural authority and expounded by eminent Acaryas, these celebrated Darsanas ensure salvation to the serious aspirant. It is well known that philosophy in India has not been merely an intellectual enterprise, but a way of life. And herein lies the Sadhanatmaka character of Indian thinking.

It is against this background that the philosophy of Yoga should be understood. While recently there is a craze for Yoga, there is less seriousness for a correct apprehension of this spiritual science. An attempt at is oversimplification and devising shortcut is what is apparent here and there. In fact, Yoga is a matter of eternal inspiration for the Sadhakas. Hence, the relevance of the repeated and reneweded studies of the basic texts of these great disciplines has dominated search of the truth since time immemorial.

In the context of such a huge and rich literature on the subject, as is amassed on Yoga to-date, it is indeed not any easy task to write a fresh commentary on the Yoga-sutras of Patanjali, which number 196, are an example of brevity and perfection with which ancient learning was preserved at the hands of Rishis.

Even an explanation of the aphoristic expression requires a mastery of the language, as well as a deeper insight into the field. While the size of the Yoga Sutras is not voluminous, it has been very rationally classified to present a scientific picture of spiritual ascent.

The first section, dealing with the nature and technique of Yoga is known as Samadhi Pada, since it is devoted to explain what the Yoga is. The second section has significantly been termed as Sadhana Pada, since it expounds the reasons for which one should adhere to the path of Yoga.

It aptly analyses the miseries of life from which liberation is sought, and it prescribes external preparatory disciplines, which attune the Sadhaka to Samadhi. The third section, called as Vibhuti Pada, lays down the internal Yogic techniques, which unfold the mysteries powers prior to attainment of Samadhi.

The last section is naturally Kavelya Pada, which thoroughly deals with the basic philosophical doctrine involved in the Yogic conception of bondage and liberation.

Sri Sant Kumar Kapoor of the Haryana Civil Service (Judicial), Palwal, has done an humble service by writing the book titled "Patanjali Unison Discipline". His modest work is more in the spirit of a Sadhaka than in that of a scholar. He has attempted a simple explanation note for each Sutra of the Patanjali Yoga-Sutras. May God grant him further inspiration in this noble direction, as the Patanjali Yoga Darsana is to first taste the ambrosia of yoga with the supreme, and then to reveal the experience. The lineage of Sadhakas as well as scholars has worked together on the Patanjali Yog Sutra from the ages to reveal and to experience the Yoga Sutras by their living evidence".

His Holiness Sri Sripad Babaji Maharaj saint of Vrindavan, the founder of Vraja Academy, Vrindavan blessed this work and made me His disciple emerging to His satisfaction of Maheshwara Sutras for proper channelization of natural order on the transcendental format of Om as Udgiti of Parvanava, AUM, Onkar and Vashitakar of absolute order of Veda:

"The revelation of Eternal wisdom has been the eternal quest—the meeting point of alpha and omega in Indian School of Thought. Lord Krishna reveals this lineage in Bhagvad Gita as:

'I taught this immortal Yoga to Vivaswan (Sun-god):

Vivaswan conveyed it to Manu (his son); and Manu imparted it to (his son) Ikswaku' IV.1

The transmission of the light of knowledge from man to man and from age to age is handed down through the great ancillary of cosmic memory by the enlightened one to the seeker in quest. Whenever this link of interaction is broken the energy is stagnant and humanity stands at the crossroad. Symbolically, the churning of divine and demonic forces comes to the point of creating problems for right discrimination to decipher the Truth and find the right dimension of consciousness.

Since these ancient links are lost, intellectual attempts to regain them are being made since several decades. The present research work within the Vedic mathematical disciplines opens new horizons for modern science and inspires rational thinking to go back to the ancient metaphysical courses of the meeting point of thingness and nothingness, which the great symbol reveal in the mantras and shruti, in the journey of consciousness of the Pauranic kathas and in the insight into the in-depth psyche of the Bhagavadgita.

Once again efforts are being made to provide a way to the rational scientist to link knowledge with the energy of an inscrutable faith, which can open in him the mystic eye or vision. The rejuvenation of this quest is deeply stressed and aims to arrive at a metamorphosis in the fields of eternal mathematics introduced here as the first of its kind.

Sri Sant Kumar Kapoor's quest – else engaged in the field of jurisprudence–has opened his intuitive faculty of study ancient scriptures in the light of modern mathematics. These studies were brought to focus by His Holiness Swami Bharti Krishna Tirthaji Maharaj, the Sankracharya of Kanchi Peeth around 1965.

As the Shrutis of the Vedas sing in chorus; "He is full, this is full, if fullness if taken out of fullness there remains fullness", so the heart of the Rigveda concludes that our thoughts should be same, our actions should be same, there should be no diversity between them, that the fullness of thought, speech and action should meet at alpha and omega, at-one-ment. To unlock the intuitive forces and dynamic energies in quest of the churned out Ambrosia the universal syllable AUM is beginning and the end.

This research work is on its way to arrive at this very conclusion. May the Supreme Divine bless this venture."

*****

For further information interested readers may visit websites:

www.vedicganita.org

www.geocities.com/vedicmathematics

www.learn-and-teach-vedic-mathematics.com

*****

More titles are available in

VEDIC METHEMATICS SERIES

1. The Teaching of Vedic Mathematics 125/-
2. Learning Vedic Mathematics on First Principles 125/-
3. Vedic Mathematics Basics 125/-
4. Vedic Mathematics Skills 125/-
5. Vedic Geometry Course 195/-
6. Learn and Teach Vedic Mathematics (PB) 125/-
7. Vedic Mathematics Space Book 195/-
8. Practice Vedic Mathematics 195/-

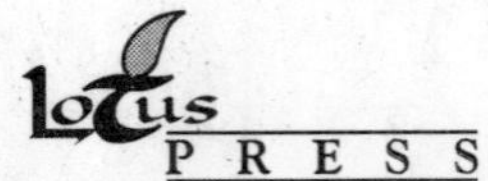

4263, Street No. 3, Ansari Road,
Daryaganj, New Delhi-110002
Ph.: 32903912, 23280047
E-mail: lotus_press@sify.com
Website: www.lotuspress.co.in